VERBS, BONES, AND BRAINS

VERBS, BONES, AND BRAINS

Interdisciplinary Perspectives on Human Nature

EDITED BY AGUSTÍN FUENTES

AND AKU VISALA

University of Notre Dame Press

Notre Dame, Indiana

University of Notre Dame Press
Notre Dame, Indiana 46556
www.undpress.nd.edu
All Rights Reserved

Published in the United States of America

Library of Congress Cataloging-in-Publication Data

Names: Fuentes, Agustín, editor. | Visala, Aku, editor.
Title: Verbs, bones, and brains : interdisciplinary perspectives on human nature /
edited by Agustín Fuentes and Aku Visala.
Description: Notre Dame, Indiana : University of Notre Dame Press, [2016] |
Includes bibliographical references and index.
Identifiers: LCCN 2016039770 (print) | LCCN 2016044814 (ebook) |
ISBN 9780268101145 (hardcover : alk. paper) | ISBN 0268101140 (hardcover :
alk. paper) | ISBN 9780268101169 (pdf) | ISBN 9780268101176 (epub)
Subjects: LCSH: Philosophical anthropology. | Human beings.
Classification: LCC BD450.V3735   2016 (print) | LCC BD450 (ebook) |
DDC 128—dc23
LC record available at https://lccn.loc.gov/2016039770

∞ *This paper meets the requirements of*
ANSI/NISO Z39.48-1992 (Permanence of Paper).

CONTENTS

Contents   vii

INTRODUCTION

The Many Faces of Human Nature

AGUSTÍN FUENTES AND AKU VISALA

The past few decades have seen an unprecedented surge of empirical and philosophical research on the evolutionary history of *Homo sapiens,* the origins of the mind/brain and human culture. This research and its popular interpretations have sparked heated debates about the nature of human beings and how knowledge about humans from the sciences and humanities should be properly understood.

To put it mildly, human nature is a contested concept. A comparison of Steven Pinker's *The Blank Slate: The Modern Denial of Human Nature* (2002) and Jesse Prinz's *Beyond Human Nature: How Culture and Experience Shape the Human Mind* (2012) shows how wildly experts disagree on the topic. For some, human nature is an enemy that needs to be abolished or an outdated scientific idea; for others, it is the cornerstone of the scientific study of humanity.

Some representatives of evolutionary psychology, for instance, fall into the latter camp. Pinker, among others, has argued that behind the dizzying variety of human cultures lies a universal psychology that severely constrains possible expressions of human thinking and behavior.

1

Furthermore, on this view, this universal psychology is largely innate in the sense that its various aspects are adaptations to the challenges that humans confronted in their Pleistocene environment. The driving force behind the evolutionary psychologist's program, it seems, is the conviction that the biological and psychological sciences can and will answer our questions about human nature.

Critics have maintained that not only are these programs based on questionable philosophical and methodological assumptions, but they also fail to account for all the relevant data. Cultures and behaviors are far too diverse to be explained by invoking an innate, universal cognition. Furthermore, recent developments in evolutionary theory seem to challenge the strong adaptationism underlying some of the arguments of Pinker-style evolutionary psychology.

One possible option is to see a much closer integration of cognition and culture in human evolution and development along the lines of niche construction or gene/culture coevolution theories. This would mean, however, that we cannot identify human nature with a set of innate cognitive mechanisms. Instead human nature should be sought from our flexible capacity to create and sustain culture and be shaped by it.

Debates revolving around innate psychological mechanisms and evolutionary psychology are not the only context in which human nature is examined and discussed. Anthropologists have long sought to identify both universal and distinct features in human cultures. Are there patterns in human cultural diversity? If a common ground among cultures can be found, perhaps that would function as a basis for shared ethical views. If there is indeed a biologically determined universal human psychology as evolutionary psychologists suggest, we should expect to see some general patterns or structures.

The quest for uniquely or distinctively human traits has also been of great interest to biologists, psychologists, and anthropologists. What, if anything, distinguishes humans from other animals? How are the psychological and biological traits that contemporary humans possess related to those of their long-dead ancestors? In addition to such scientifically motivated questions, there are a host of perennial philosophical, ethical, and theological issues. Indeed, before the emergence of modern biology, psychology, and social science, philosophy and theology were the main sources of reflection about what humans are and what they are like. Questions concerned the nature of the human soul, virtue, emotions, and free will, as well as morality. Are we naturally inclined toward altruism and selflessness as some classical liberal thinkers such as Rousseau claimed, or is our state of nature war against everybody as Hobbes insisted?

Even from these brief reflections, it is clear that "human nature" has a strong ethical and political component. Whether there is such a thing as human nature can surely make a difference in the way we live our lives. In our contemporary situation, we are faced with controversies over the possibility of human enhancement, the ethical challenges of biological technologies, the beginning and end of life, and various expressions of human sexuality. All these debates involve assumptions about what humans are like, whether they have some natural goals or ends.

It is abundantly clear that there is no single discourse or debate about human nature. By the same token, there are various distinct and more or less incompatible quests for human nature, and it is often unclear where the boundaries between these various enterprises should be drawn. This is the reason we recommend taking a wide, interdisciplinary stance when discussing human nature. Such an interdisciplinary conversation should not be restricted to the sciences alone. Instead, we should take into account perspectives from all academic fields, including philosophy, theology, and the humanities as a whole.

Such problems formed the background for Agustín Fuentes's Human Nature(s) Project: Assessing and Understanding Transdisciplinary Approaches to Culture, Biology and Human Uniqueness (2011–14). The John Templeton Foundation funded the project, whose aim was to provide a road map of how human nature is approached in different fields, from anthropology, philosophy, and theology to biology and psychology. The project culminated in an interdisciplinary conference held at the University of Notre Dame in April 2014. The aim of the conference was to address the various issues associated with human nature in the different disciplines and focus on deeper integration between them. This volume presents the papers and responses from the conference, offering a state-of-the-art look at the debates about human nature.

Both the conference and this edited volume have taken a consciously transdisciplinary approach. Often, the focus is on a single perspective, usually in the context of a specific science, like biology or neuroscience. We hold, however, that what it means to be human cannot be reduced to a single perspective or definition, scientific or otherwise, which is why we have included perspectives from theology, philosophy, and the social sciences, as well as the sciences. This is important because our various inquiries into human nature are closely connected to ultimate questions regarding the nature of the cosmos, human origins, teleology, and the ways in which we obtain knowledge.

In this introduction, we offer a brief overview of the human nature debates so far, make some basic distinctions, and identify issues of agreement and disagreement among most scholars involved.

Quests for Human Nature

As we pointed out above, human nature has been a perennial philosophical and theological topic. If one looks at, for example, Louis Pojman's *Who Are We* (2005), which offers an overview of the philosophical debates in the Western tradition, one finds a plethora of topics, including the following:

> Do humans have souls, and if so, what are those souls like?
> What is human psychology like?
> What are the essential features of human beings as opposed to animals?
> Are we free, and what is the nature of our freedom?
> Are we naturally good or evil, selfless or selfish?
> Where did we come from, and what is our ultimate end?
> What kind of society is good for us?
> What should our proper relationship to God be?

In the contemporary scene, these topics are often discussed separately in respective branches of psychology, political theory, philosophy, and theology. In this context, it is futile to try to summarize the vast literature on all these topics. It is enough to point out that reflection on these questions is still very much part of at least some discourses, albeit more philosophical than scientific, about human nature.

In many cases, however, philosophical discussions have gravitated toward a rather robust notion of human nature—what we might call the *everyday* or *folk* concept of human nature. It seems that our folk concept of human nature is a source of much confusion, because of the various aspects of it that in principle could come apart.

Common notions of human nature assume a specific essence to being human, a kind of transhistorical core for all human beings. If this were true, underlying all human diversity and variation there would be something constant, some traits, tendencies, and capacities that all humans share. Usually this is cashed out in terms of distinguishing human biology from the influence of culture and upbringing: we have the same genes on top of which culture and upbringing assert their influence.

This folk concept of human nature seems to entail that the essential and universal human features have the same causal history. That is to say, there is something very deep in human biology or some sort of a plan in nature itself that causes the similarities. Traditionally, such causal powers were usually attributed to something like the soul or the self inside the individual or the purposes of God or the brute purposefulness of nature. It is the innate essence that is causally responsible for our uniquely and universally human features.

The third aspect of the folk concept of human nature is a normative one: given that there is a purposeful human essence, some human actions, practices, or societies are more natural than others. Unnatural is not only bad but also wrong, whereas natural is good and right.[1]

With Charles Darwin's writings, a new perspective on human nature emerged. Darwin was the first to offer systematic tools and theories linking human capacities, development, and history to that of other animals: the same kinds of causes that are driving change in other animals also work in the human case. This put the whole question of human nature in a new light. The move from an essentialist notion of species to an evolutionary, Darwinian notion of species as populations seemed to cast doubt on there being an innate essence to humanity. The Darwinian turn also seemed to undermine the teleological or normative aspects of the folk notion of human nature. Natural selection, Darwin maintained, does not work with a goal or end in mind.

At the same time that the Darwinian turn created pressure to see human nature in biological, evolutionary terms, the social sciences and humanities were going in a completely different direction. Nineteenth- and twentieth-century European thinking in philosophy and social theory had a strong tendency to reject both the everyday and Darwinian notions of human nature. Marxists and existentialists, for instance, maintained that there is no such thing as a biologically given essence. For the Marxist, it is our institutions and relationships that make us what we are, and society is the sum of these. However, these relationships are in no way given by nature but can be reorganized. Thus by changing society, we change human nature. For the existentialist, it is not biology or society that determines our nature but our own will. We are thrown into the world and without the help of God or nature must decide what we want to be. Both Marxism and existentialism, thus, point to a kind of anti-essentialism about human nature.

Although Darwinian ideas were developed after Darwin's death, the evolutionary perspective remained a minority view in the social sciences. Controversies began in the 1970s when the ethologist Edward O. Wilson introduced the idea of *sociobiology*. Working inside the newly formed Modern Synthesis, Wilson adopted an uncompromisingly neo-Darwinian view of human social behavior, especially altruism: outwardly altruistic behavior can be explained by invoking the underlying fitness effects it has for individuals. In other words, human social behavior is not to be explained by cultural influences or some such factors but instead by seeing it as an adaptation for survival and reproduction. Here Wilson envisaged a program under which the social sciences and psychology would ultimately be subsumed under evolutionary biology. In terms of human nature, Wilson's program amounted to a rather robust view: under the veneer of individual and cultural variation, humans share a set of biological dispositions and traits that is not too far from our primate cousins. So not only is there a biologically driven innate human nature, but there is no unique or distinct human nature to speak of.

Wilson's claims created a large-scale debate on various overlapping issues, sometimes known as the sociobiology debate. Some of Wilson's biologist opponents, like Stephen Jay Gould, challenged his heavy reliance on natural selection as the only possible explanation for traits of organisms. For many social scientists at the time, the explanations for human behavior were sought at the level of cultures, institutions, and individual motivations. Wilson turned this upside down: now it was our genetic inheritance shaped by our evolutionary past that held cultures, institutions, and motivations on a tight leash. Finally, the sociobiology debates would have not been so vitriolic were it not for the ethical and political implications of Wilson's work. Many of his opponents saw him as offering a pseudoscientific defense of a rather conservative or traditionalist view of human nature in which human beings are naturally selfish.[2]

In the 1980s, sociobiology debates coupled with the advance of cognitive science and developmental cognitive psychology gave birth to the diverse field now known as evolutionary psychology. Evolutionary psychologists maintain, against sociobiology, that we should not jump directly from the possible adaptive effects of social behaviors to the explanation of current behaviors. Instead, we should focus on the mechanisms of the human mind that are both products of natural selection and causes of current behaviors.

We want to point out that there is no single "evolutionary psychology." David Buller suggests that we should distinguish the standard paradigm of evolutionary psychology from the general field of evolutionary psychology. Evolutionary psychology as a field of research is a broad and loosely connected group of diverse disciplines that includes behavioral ecology, human ethology, and evolutionary anthropology. In this sense, "evolutionary psychology" is a general term for approaches that use contemporary evolutionary theory as background for psychological, anthropological, or neuroscientific theorizing.[3]

However, the standard model (or nativistic model, or the "Santa Barbara school") of evolutionary psychology refers to a more specific approach represented by Pinker, David Buss, and Leda Cosmides and the anthropologist John Tooby, among others. The central idea of this approach is that the human mind is a massively modular system whose modules were created by natural selection to solve adaptive problems in our ancestral environment. Now these once adaptive and panhuman modules are used for different tasks, and they form the universal basis of our diverse cultural forms. Under the veneer of contemporary culture, we all still have Stone Age minds.[4]

Again, the standard or nativistic model of evolutionary psychology leads to a rather robust view of human nature. It is at this level of cognitive mechanisms where human nature is found. For Cosmides and Tooby, the modules of the mind are "the psychological universals that constitute human nature."[5] Regardless of culture, humans develop a rather unified set of psychological mechanisms, so Cosmides and Tooby can also claim that "human nature is everywhere the same."[6] This nature is innate in the sense that it is genetically coded and invariably produced by normal human development.

It is worth pointing out that sociobiology and evolutionary psychology and their opponents are mainly interested in explaining human behavior. The controversy is ultimately about where to look for explanations of human behavior: innate biology or psychology or culture and experience. If there are kinds of human behavior that have pretty much the same biological and psychological causes regardless of cultural context, then a theory of human nature is viable. If there is no innate human nature and there are no general, scientifically tractable facts about human behavior (because human behavior is, say, so context-sensitive), the scientific quest for human nature does not seem like a worthwhile enterprise.

Diverse fields of evolutionary human sciences have emerged in the wake of evolutionary psychology and related debates. By the 2010s, the variety of theories and approaches was dizzying: Fuentes's *Evolution of Human Behavior* (2009) catalogs over thirty theories. The diversity of topics and issues can also be seen in a new collection of classic and new articles, *Arguing about Human Nature* (edited by Stephen Downes and Edouard Machery). Many topics have their roots in the sociobiology debates, but they also go beyond it. For example:

Evolution of human nature, that is, whether biology can offer a robust view of human nature.

The origin of the human mind and its fixity or plasticity. Is it the case, for instance, that our minds were formed and fixed during the Pleistocene period, or did human minds change after that? Can they change now?

Issues having to do with innateness, that is, whether we have innate psychological mechanisms or traits and what this innateness might mean.

Claims about genetic determinism and what the sciences say about the relationship between genes and human behavior.

Universality of human traits and issues having to with individual variation.

The nature of cultural and social categories, that is, whether these categories are natural kinds or socially constructed.

The question of human genetic diversity and the concept of race.

Debates about the extent to which human sexual behavior can be explained in evolutionary terms.

Numerous political and ethical issues revolving around health and the notions of "normality" and "species-typical functioning."

Common Ground on Human Nature

There are some issues on which many scholars in this volume find themselves in modest agreement. Three of these issues are outlined briefly below: (1) commitment to interdisciplinarity, (2) rejection of robust forms of evolutionary psychology and sociobiology, while still maintaining that (3) human nature might be a relevant or useful notion.

Commitment to Interdisciplinarity

There is a widespread commitment to interdisciplinarity and openness to various fields of study (see Fuentes and Visala 2016). This is something that the project, the conference, and this volume have sought to both foster and push further. Indeed, in a recent article, one of us (Agustín Fuentes), following the lead of the theologian J. Wentzel van Huyssteen, outlines a transdisciplinary approach to the study of humans (Fuentes 2013b; see also van Huyssteen 1998, 1999, 2006).[7] Given the multifaceted nature of human beings, Fuentes argues that interdisciplinarity and multidisciplinarity are not enough. While not tightly defining the concepts, Fuentes sees *interdisciplinarity* as collaboration across disciplines and *multidisciplinarity* as utilizing two or more disciplines to study the same subject matter. But for the purposes of studying human nature, neither is enough.

Both van Huyssteen and Fuentes seek a deeper engagement between disciplines: on what they call *transdisciplinarity*, not only do disciplines share some aspects of the subject matter, but they end up, potentially, changing one another—perhaps adopting new methods, questions, and approaches from another discipline. In other words, the goal of transdisciplinarity is to change the disciplines involved by influencing the methods, worldviews, and languages used in each. This change goes deeper than just collaboration on similar subject matter, as in interdisciplinarity and multidisciplinarity.

According to Fuentes, such transdisciplinary approaches are particularly useful in broad or highly integrated topics that have previously been the purview of strongly "foundationalist or positivist traditions" (Fuentes 2013b, 108–9). This argument posits that transdisciplinary work is needed in order to make sure that each discipline takes into account and has access to all relevant data about human behavior and evolutionary history. Such successful transdisciplinarity should also lead to synthesizing knowledge across disciplines in order to obtain a coherent view of the subject matter.[8]

Van Huyssteen is somewhat more critical of the latter goal. He suggests that we need to get away from seeing human nature from a single perspective and adopt a critical stance toward all disciplines and approaches that try to dominate others or present themselves as the ultimate one. His *postfoundationalist* methodology entails that different epistemic traditions or disciplines have different rationalities. Rationality, for van Huyssteen, is an embodied cultural practice of answering the questions the tradition seeks to answer. In terms of content and method, however, traditions overlap. These overlaps van Huyssteen calls *transversal points* or spaces. Interdisciplinary work can be done in these transversal points, and they can be expanded, but we cannot demand that other traditions accept our rationality or worldview. In this sense, the postfoundationalist approach wants to affirm that large-scale intellectual traditions are "safe," in the sense of not being under threat of falsification, reduction, or elimination from others. Work at transversal points, however, can (and should) change each tradition that is involved in the exchange. Nevertheless, the ultimate aim is not to fuse traditions (or disciplines) or eliminate them but to make them develop and change.

As Fuentes's and van Huyssteen's approach suggests, commitment to interdisciplinarity involves (or should involve) rejecting two methodological extremes about the relationship between the sciences and humanities. On the one hand, there are those who draw a sharp distinction between the sciences and the humanities and claim that all quests for human nature are just expressions of scientific imperialism or reductionism. On this view, human thinking and behavior are not the proper subjects of the sciences: since there are no laws of thought and behavior, we should seek to understand human beings, as it were, from the inside.

On the other hand, there are those who think that the humanities have very little to contribute to discussions about human nature. The humanities and social sciences might have identified some structures or patterns of human cultures and behaviors, but the causes of these can be found in either universal human psychology or genetic inheritance. In other words, the humanities and social sciences do not provide explanatory theories but simply raw data for the sciences to explain. Ultimately the social sciences and humanities will be integrated into psychology and biology, forming a unified science of the human. It seems that strong forms of both sociobiology and evolutionary psychology either assume or imply something like this.

Against these two trends, most participants in contemporary human nature debates share a common in-between ground. The details vary from thinker to thinker, of course, but the basic idea might be something like the following. First, there is no a priori or strict distinction between "understanding" and causal explanation. Thus the natural sciences, humanities, and social sciences are in the business of explaining human culture, behavior, and development in some broad sense. It seems to us that the strict distinction between understanding and explanation was based on an outdated view of explanation as a search for universal laws—a view that has fallen into significant disrepute since the 1970s, at least among philosophers of science.

One possible way to give content to this middle view would be critical scientific realism (e.g., Niiniluoto 1999; Psillos 1999). For critical realists, all academic disciplines seek knowledge of (often) unobserved aspects of our world; for the critical realist, therefore, the "world" is not restricted to what we can directly observe but can include nonobservable, causally efficacious entities. Nevertheless, knowledge of this objective world is something that we seek. Critical realists would also reject strong forms of social constructivism and cultural relativism to the extent that they either deny or disregard the possibility of real knowledge about our world. Finally, critical realists hold that our knowledge is always subject to revision when new evidence comes in.[9]

Not from Genes Alone

We also see a consensus forming about what we should think about strong forms of sociobiology and evolutionary psychology. Although these points of view have contributed significantly to the conversation on human nature, most authors included in this book would want to steer clear of strong notions of innate or otherwise biologically determined human nature. There are various reasons for this. We briefly mention two: recent developments in evolutionary theory and doubts about the fixity of human cognition.

First, modern evolutionary theory is not entirely reliant on explanations that depend exclusively on natural and sexual selection: the whole of human evolutionary experience is not reducible to variation in patterns of DNA and reproductive fitness. Against this, sociobiology and much evolutionary psychology still rely wholly on natural (and sexual) selection as the only significant architects of evolutionary change. That modern evolutionary theory has moved substantially beyond this perspective is now challenging the evolutionary assumptions of both sociobiology and evolutionary psychology. Our contemporary view is that evolution goes way beyond "survival of the fittest."

Current understanding of evolution can be summarized as follows: mutation introduces genetic variation, which, in interaction with genetic drift and epigenetic and developmental processes, produces biological variation in organisms that can be passed from generation to generation. Gene flow moves the genetic variation around, and natural (and sexual) selection shape variation in response to specific constraints and pressures in the environment. Organism-environment interaction can result in niche construction, changing the shape of natural selection and creating ecological inheritance. For humans, social structures/institutions, cultural patterns, behavioral actions, and perceptions can affect these evolutionary processes. This in turn can affect developmental outcomes. Diverse systems of inheritance (genetic, epigenetic, behavioral, and symbolic)[10] can all provide information that influences biological change over time.

In addition, the emergence of epigenetics has significantly changed our view of the relationship between evolution and the development of an organism.[11] Epigenetic processes can affect gene function and regulation but are not coded for in DNA. These processes are initiated, regulated, and otherwise influenced by life experience, social stressors, perceptions, and a range of psychological variables, in addition to being affected by specific biotic and material ecological factors and having cross-generational impacts. Thus epigenetic variations can produce different outcomes even for organisms with identical DNA sequences.

Humans generate and transmit symbols, artifacts, institutions, and meaning, in addition to our ecological manipulations and genes. All of these processes are multidirectional, with humans, throughout their life spans, both directing and being directed by their own development (Flynn et al. 2013; Kendal 2012). Human reliance on learning, plasticity, and culture lends human niche construction a special potency. Niche construction includes the effects of the cultural context, social histories, and human behavior as an active part of our evolutionary dynamic.

These emerging perspectives put the relationship of evolution, development, and culture in a new light. It has become clear that human biology does not exist separate from our social and structural ecologies: our culture and cognition are constantly entangled with our biology (Fuentes 2013a). The boundaries between our genes, epigenetic systems, bodies, ecologies, psychologies, societies, and histories are fluid and dynamic. Perception, meaning, and experience are as central in our evolutionary processes as are nutrients, hormones, and bone density—and all these elements can interact.

We want to draw attention to another strand of criticism of standard or nativistic evolutionary psychology. Many critics have challenged the underlying idea of rigid, genetically specified cognitive structure. Against this, David Buller (2005), among others, argues that a rigid psychology would not be what natural selection would most likely produce. He writes:

> Even if our species was faced with recurring adaptive problems throughout a significant portion of its evolutionary history, distinct "genetically specified" brain circuits were not required to solve those

problems. Our brains hit on a different, domain-general solution: a plasticity that allows particular environmental demands to partici-pate heavily in tailoring the cortical circuits that process information about those demands. . . . We didn't evolve a separate, genetically specified brain circuit for each adaptive-problem domain encoun-tered in our evolutionary history. (140)

The human ability to learn is grounded in the plasticity of the brain's sys-tems. It is much more likely that evolution would favor a few flexible and malleable mechanisms rather one fixed structure.

Eliminativism about Human Nature

The third issue where there is at least some agreement is that the notion of human nature might be useful or relevant and should not be rejected completely. One reason for skepticism or eliminativism about human na-ture might be, as we have already seen, a strong methodological commit-ment to separating the sciences from the humanities. Related to this, radical forms of social constructivism or cultural relativism might also criticize the notion of human nature on ethical or political grounds: claims about essential human nature can be used as tools in the hands of oppressors.

There are also various scientific arguments for eliminativism about human nature. One commonly used argument is that of David Hull and others, which maintains that since Darwinian population thinking elimi-nates species essences, it makes no sense to talk about "human" or "human nature" from a biological point of view. Another scientific argument has to do with criticisms of strong forms of evolutionary psychology. As we have seen, some representatives of evolutionary psychology, like Pinker, maintain that there are robustly innate cognitive mechanisms that are the basis of human nature. Critics of this line of thought might argue that because there are no robustly innate and specified mental mechanisms, there is no human nature in any meaningful sense. In other words, there is no human nature, because there is no biologically or genetically given psychology. Such an argument is, more or less, present in several chapters in this volume. We could dub the former argument anti-essentialist from Darwinism and the latter the nurturist argument.[12]

A burgeoning philosophical discussion has developed about the prospects of some scientifically relevant human nature. Grant Ramsey's chapter in this volume is a contribution to that discussion. Ramsey and others, including Edouard Machery and Tim Lewens, seek to formulate notions of human nature that are knocked out by neither the anti-essentialist nor the nurturist argument. Indeed, Ramsey and Machery explicitly reject various forms of innateness and associated ideas of genetically specified cognitive mechanisms.

It is also worth pointing out that even if the scientific quest for human nature ends up being futile, who is to say that there cannot be an account of human nature that is not scientific. Indeed, many a philosopher (and theologian) might say that there is an essential human nature, but it is not accessible in any direct way by science. One could, for instance, claim that humans are essentially self-conscious persons with a certain kind of moral standing.

MOST OF THE AUTHORS in this volume seek a meaningful middle way between strong nativism and eliminativism about human nature. They believe that there could be a meaningful use for the concept "human nature." We believe that by exploring the interfaces of cultural and biological evolution, human nature and personality, and the flexibility of the human mind such a middle way can be found.

Dividing Issues

Now that we have some bases for common ground, let us outline three topics on which there is considerable disagreement among most participants: what it means to have a nature, what we are, and whether or not we should accept some form of naturalism.

What It Means to Have a Nature

It seems to us that the most fundamental disagreement in the debates about human nature is what it means for us to have a nature in the first place. The issue is complicated because the word *nature* can be understood in several different ways. Nature and its cognates in what sense:

nature as opposed to culture, natural as opposed to unnatural, natural versus accidental, and so on? Furthermore, "human nature" as a concept has considerable philosophical, ethical, and political baggage, as we have seen.

We already outlined the everyday concept of human nature, how it was appropriated by sociobiology and evolutionary psychology and how it was subsequently criticized. The folk notion and many of its incarnations in evolutionary psychology contain various separate components. For the sake of clarity, we should distinguish them.

The attempt to reveal our *universal human nature* has to do with, simply put, human universals. What are the invariant dispositions of human beings? Perhaps there are anatomical or physiological or psychological or social characteristics that all human individuals or societies share. If there are such features, then they constitute our nature. Candidates for human universals include language, music, marriage, play, and various other cultural institutions. The quest for psychological universals that motivates cross-cultural and anthropological research like the work of Donald Brown is an example. Many take this quest to be of great ethical and political importance, seeing commonality among human beings as the basis of human rights, for example.

The second notion of human nature is that of *human uniqueness*. What is it distinguishes human beings from other things, including other animals? Is there, perhaps, a set of properties that not just all but only human beings share? Biologists might be interested in this question for taxonomic purposes; theologians might wonder what the *imago Dei* consists in. Many candidates for human uniqueness have been offered: language, rationality, self-consciousness, capacity for moral judgment, religion.

The problem is that we do not really have an idea of what exactly uniqueness means. First of all, there is the uniqueness of certain human abilities. I doubt that anyone would deny that humans can do a lot of things that no other animal can do. We can build space shuttles and satellites; write books about human nature and design computers. What is controversial is whether such abilities are based on unique capacities. In other words, do humans have capacities that other animals do not have? It is clear that we share most of our psychological capacities pretty much as they are with many other animals (sight, hearing, basic forms of reasoning, etc.). But do we have capacities that significantly differ from those of other animals? Traditionally, candidates for such capacities have included higher-level reasoning, the intellect, moral judgment, symbolic language, and knowledge of God.

Finally, one could say that humans are unique in the sense that they are not products of the same causes as other animals. Here the uniqueness would be at the level of causes or factors that contribute to the emergence of humans. If it is the case that humans are products of the same causes as all other animals, then there is no human uniqueness in this sense. But if there are causes that contribute specifically to human evolution, say, gene-culture coevolution or kinds of niche construction that are not present in other animals, then we might be unique in this sense.

Note that in all these cases uniqueness is a matter of degree, and it is extremely difficult to draw sharp boundaries. So we think that we should be extremely critical of sweeping claims about similarities or dissimilarities between humans and other animals.

Then there is the idea of *innate human nature*. The fixity of traits across cultures has typically led to questions about whether such traits are "innate." The idea is that the traits that constitute human nature are somehow produced by innate tendencies or causes (genetically specified or other). However, it is far from clear what "innateness" consists in. A trait may be "innate" in the etymologically conservative sense that it is present at birth, for example. This is closely related to the notion that innate traits are those that are not learned. If—and this is a big if—it makes sense to assign causality between genetic and environmental variables, an innate characteristic might be one that is wholly or mostly due to the genetic influence. We could also claim that a trait is innate if it is "environmentally canalized" or "canalized with respect to the environment," that is, if its development is insensitive to environmental variation. Candidates for innate nature consist again of the usual suspects: language, morality, religion, and reasoning. Many debates about human nature are actually about innateness, as we can see from, for example, Pinker and Prinz.

We should also remember that notions of human nature could include normative components. Typically, the folk concept of human nature has normative implications: natural behaviors and tendencies are good; non-natural are bad. A further normative claim would be that whatever powers and capacities human nature consists of, its realization would be something morally right. If we naturally act selfishly, this would be the right way for us to behave. Similarly, if we were naturally selfless, this would be the right way for us to act. Traditionally, some ethical theories, like natural law ethics, have drawn extensively on what they have considered to be "natural" for humans and in the process produced various definitions of human nature. Other ethical theories, such as deontological ethics, have opposed this by arguing for a strong distinction between nature and what is normative.

Typically, when people argue that there is no human nature or that it is not sufficiently robust, they have only one possible notion of human nature in mind. Jonathan Marks's chapter in this volume is a good example. He argues that we need to stop thinking about human nature as opposed to culture. Instead, human evolution has taken place in the context of culture and human sociality, so there is no way to clearly distinguish a part of human behavior produced by "nature" from the behavior produced by "culture." We are biocultural animals. While Marks draws attention to the normative aspects of human nature concepts, his arguments against innate nature say very little or nothing about unique or universal human nature.

What Are We?

Not only is the "nature" part of human nature problematic, but so is the "human" part. This might seem somewhat counterintuitive at first, so let us explain. We take it for granted that whatever quest for human nature we embark on we are ourselves the subject of that quest. In other words, the "human" in human nature refers to us, not to some obscure beings across the galaxy. We are the ones whose nature is in question. However, it is far from clear what we are, what kinds of beings we are. What counts as a human being?

There seem to be two different starting points here, both supported by a basic set of human intuitions. On the one hand, we identify ourselves as a specific kind of animal, *Homo sapiens*. In this case, the quest for human nature is the quest for the psychological, social, or biological features of our species. In contemporary philosophical literature, this view is often called animalism. As a starting point for a theory of human nature, animalism has a problem. This has to do with the fluidity of Darwinian species concepts discussed above. We might be identical to *Homo sapiens*, but there seems to be no clear way of distinguishing members of *Homo sapiens* from other species. It is easy, of course, to distinguish currently living *Homo sapiens* from chimpanzees, for instance, but this does not mean that there is an essential set of properties or traits that would define the category *Homo sapiens*. Surprisingly, animalism has not been very popular in the Western philosophical tradition. Indeed, it has been more typical to think of "us" as animals only accidentally (i.e., nonessentially). That is, while it is true that we are animals—by virtue of the bodies we have, the behaviors we exhibit, and so on—we are, philosophers and theologians have historically maintained, essentially persons, thinking things.

There have been various options along these lines, but we mention only two: dualism and constitutionalism. On the dualist view, a human individual is identical with a nonphysical soul. We only accidentally have bodies and *Homo sapiens* species membership; we could have different kinds of bodies altogether, non–*Homo sapiens* bodies. Some forms of physicalism also maintain that we are essentially persons and that human personhood is not just membership in the species *Homo sapiens*. According to constitutionalism, what makes us persons is that we have a first-person perspective; that is, we are self-conscious and can refer to ourselves with the first-person pronoun "I." Further, to have this capacity for first-person perspective is essential for us; we would not exist without it. Nevertheless, I am not identical with a nonphysical soul, nor am I identical with a particular *Homo sapiens*. Rather, the *Homo sapiens* with which I share all my parts constitutes me.

In the volume at hand, the philosopher Carl Gillett defends an alternative to dualism, constitutionalism, and animalism. On his view, we are identical to a specific part of *Homo sapiens*, the brain. The brain view, as Gillett points out, resembles dualism and constitutionalism in the sense that it takes the mental as essential for our existence: we are things that think. Animalism, however, does not do this: our existence as animals does not essentially involve thinking or mentality. The brain view differs from dualism and constitutionalism in the crucial respect that instead of attributing thinking to the soul or to the first-person perspective, it attributes thinking to the brain. Since we are thinking things and the brain is the part of the *Homo sapiens* that thinks, the brain view claims that we are identical with brains.

Should We Commit to Naturalism?

Finally, we want to highlight a topic of disagreement that has to do with broad metaphysical and epistemological views. Both Fuentes and van Huyssteen point out that what we think about metaphysics and epistemology in general might have important implications for our view of human nature. This is nowhere more evident than the debates between religious, theological, and naturalistic views of human nature. For theists and non-naturalists, there might be aspects of human nature that are not scientifically tractable. Some theologians have suggested that only revelation tells us what human beings ultimately are. Scientific study of humans might be useful in many ways, but it does not, on this view, tell us much about our essential nature. Theologians, however, are not the only non-naturalists. Some philosophers, who identify themselves as naturalists, also maintain that certain aspects of human nature, such as self-consciousness, reason, and freedom, might be inaccessible to science (e.g., Nagel 2012).

It seems to us that many debates about human nature exhibit a kind of naturalistic bias. There is an implicit assumption that scientific results and debates about human nature are free of philosophical or conceptual assumptions. The truth of animalism, for instance, is often assumed without any further argument. Similarly, it is often assumed that whatever human nature is, it is scientifically and empirically tractable, making a theory of human nature naturalistic. We are not trying to say that these views are necessarily false and that naturalism must be false as well. Instead, what we are saying is that these assumptions need to made explicit and argued for. The naturalist cannot come to the transdisciplinary table to discuss human nature and simply assume that everyone else will agree with her on these issues.

Our background assumptions also have an implicit or explicit impact on how we understand the questions: biological accounts of how *Homo sapiens* emerged, for instance, do not necessarily answer all our questions about what human nature means. We do not want to say that there is a necessary contradiction between asking biological questions and asking other kinds of questions about human beings. Instead, we want to point out that posing the questions biologically might end up obscuring or completely sidelining all other ways of addressing and understanding what humanity is and what its nature is like.

Some chapters in this volume are clearly committed to a, broadly speaking, naturalistic stance. Ramsey, for instance, argues that human nature, whatever we mean by it, has to be a scientifically and empirically tractable notion. Jonathan Marks argues for the rejection of somehow purely genetically or biologically grounded human nature. There are also writers, such as Neil Arner and J. Wentzel van Huyssteen, who defend a broader stance, one that could include nonscientific sources of knowledge.

Our goal here is not to enter or adjudicate complex philosophical and worldview issues of an epistemological and metaphysical nature. Instead, we want to make two points: first, there is widespread (and in our view, rational) disagreement about basic metaphysical and epistemological questions (the existence of God, the nature of scientific knowledge, existence of essences, etc.); second, we must be cautious and try to explicate and identify where our own assumptions come in when we talk about human nature. Nevertheless, we want to emphasize that we should not give up seeking convergence on human nature because of our metaphysical disagreements. If convergence eludes us and common ground cannot be found, this is already a relevant result: it would reveal to us how deeply our questions about ourselves are connected to our basic metaphysical assumptions.

A Preview of Coming Attractions

The rest of the book is divided into chapters, each of which consists of a main essay followed by responses to and comments on its themes. The first two main essays argue against robust, biological determinants of human nature. In the first chapter, the anthropologist Jonathan Marks offers a broad set of arguments as to why we should reject the opposition between biology and culture in shaping human nature. We have misunderstood the nature of human evolution if we imagine that we can somehow isolate a biologically given, invariant human nature from human culture and sociality. For Marks, such attempts also reveal how science is often used to drive various ethical and political agendas.

In the second chapter, the anthropologist Tim Ingold outlines a new approach to thinking about human nature. For Ingold, there is no innate, given human nature. Instead he talks about human becoming and how our development is constantly shaped by close interactions with the environment. The conclusion of both Marks and Ingold is that human nature is not to be found in biology or psychology somehow separated from human culture and social life: a much broader perspective that includes culture, evolution, and learning has to be taken into account.

The next three chapters seek to provide accounts of what human nature means and how we might give robust content to these notions without falling into the traps that both Ingold and Marks describe. In their "Recognizing the Complexity of Personhood: Complex Emergent Developmental Linguistic Relational Neurophysiologicalism," the neuroscientists Warren Brown and Brad D. Strawn describe the ways in which multiple brain mechanisms interact with our environments during human development and give rise to distinctively human capacities like language and complex cognition.

The theologian J. Wentzel van Huyssteen explores how theologians could (and should) take into account human evolution and human prehistory when they reflect on human nature. For van Huyssteen, the archaeological work on Catalhoyuk is especially informative and tracks the emergence of the human sense of self in prehistory.

Finally, the philosopher Grant Ramsey provides a window onto current debates about human nature in the philosophy of science. After outlining some reasons for pessimism about human nature, he develops a notion of it that could function as a starting point for empirical research.

In the spirit of open, transdisciplinary dialogue we invited the respondents to address some themes or issues in the main essays, without having to respond to the main essay point by point. As a result, some commentaries and responses are direct engagements with the main essays, whereas others are more like short presentations about a related topic.

The book closes with an epilogue consisting of two reflections about how to proceed in our (various) quest(s) for human nature. The focus here is how the interdisciplinary, and possibly transdisciplinary, conversation can be enhanced and where a common ground might be pushed further and what the dividing issues are.

Notes

1. A very interesting empirical study on the folk concept of human nature is Linquist et al. 2011.

2. Segerstråle 2002 is a great introduction to the sociobiology debate.

3. For an overview, see Barrett, Lycett, and Dunbar 2002.

4. The basic assumptions of evolutionary psychology are: (1) computational theory of mind, (2) strong nativism, (3) adaptationism, and (4) massive modularity. The seminal work of this group is Barkow, Cosmides, and Tooby 1992; Buss 1999 is its new "bible." For criticisms of evolutionary psychology, see Buller 2005; Richardson 2007.

5. Quoted in Buller 2005, 71.

6. Quoted in Buller 2005, 71.

7. See also Green 2010.

8. Hadorn et al. 2008; Kessel and Rosenfeld 2008; McDonell 2000.

9. One application of critical realism to the social sciences is Smith 2010.

10. Jablonka and Lamb 2005; see also Laland et al. 2014.

11. Epigenetics is the study of interactions in development above the level of the gene.

12. For more details, see Visala's response to Grant Ramsey in this volume.

References

Barkow, J. H., L. Cosmides, and J. Tooby. 1992. *The Adapted Mind: Evolutionary Psychology and the Generation of Culture.* New York: Oxford University Press.

Barret, L., J. Lycett, and R. I. M. Dunbar. 2002. *Human Evolutionary Psychology.* Princeton, NJ: Princeton University Press.

Buller, D. J. 2005. *Adapting Minds: Evolutionary Psychology and the Persistent Quest for Human Nature.* Cambridge, MA: MIT Press.

Buss, D. M. 1999. "Evolutionary Psychology: A New Paradigm for Psychological Science." In *Evolution of the Psyche,* ed. D. H. Rosen and M. C. Luebbert, 1–33. Westport, CT: Praeger.

Downes, S., and E. Machery. 2013. *Arguing about Human Nature: Contemporary Debates.* London: Routledge.

Flynn, Emma G., Kevin N. Laland, Rachel L. Kendal, and Jeremy R. Kendal. 2013. "Developmental Niche Construction." *Developmental Science* 16 (2): 296–313.

Fuentes, A. 1999. *Evolution of Human Behavior*. Oxford: Oxford University Press.

———. 2013a. "Blurring the Biological and Social in Human Becomings." In *Biosocial Becomings: Integrating Social and Biological Anthropology*, ed. T. Ingold and G. Paalson, 42–58. Cambridge: Cambridge University Press.

———. 2013b. "Evolutionary Perspectives and Transdisciplinary Intersections: A Roadmap to Generative Areas of Overlap in Discussing Human Nature." *Theology and Science* 11 (2): 106–29.

Fuentes, A., and A. Visala. 2016. *Conversations on Human Nature*. New York: Left Coast Press/Routledge.

Green, E. 2010. "A Primer in Interdisciplinarity: J. Wentzel van Huyssteen and the Postfoundational Approach." *Toronto Journal of Theology* 27 (1): 27–36.

Hadorn, G. H., S. Biber-Klemm, W. Grossenbacher-Mansuy, H. Hoffmann-Riem, D. Joye, C. Pohl, U. Wiesmann, and E. Zemp. 2008. "The Emergence of Transdisciplinarity as a Form of Research." In *Handbook of Transdisciplinary Research*, ed. G. H. Hadorn et al., 19–42. Zurich: Springer.

Jablonka, E., and M. Lamb. 2005. *Evolution in Four Dimensions: Genetic, Epigenetic, Behavioral, and Symbolic Variation in the History of Life*. Cambridge, MA: MIT Press.

Kendal, J. 2012. "Cultural Niche Construction and Human Learning Environments: Investigating Sociocultural Perspectives." *Biological Theory* 6 (3): 241–50.

Kessel, F., and P. L. Rosenfeld. 2008. "Toward Transdisciplinary Research: Historical and Contemporary Perspectives." *American Journal of Preventive Medicine* 35 (2S): S225–S234.

Laland, K. N., T. Uller, M. Feldman, K. Sterelny, G. Muller, A. Moczek, E. Jabonka, and J. Odling-Smee. 2014. "Does Evolutionary Theory Need a Rethink? Yes, Urgently." *Nature* 514: 161–64.

Linquist, S., E. Machery, P. Griffiths, and K. Stotz. 2011. "Exploring the Folk-biological Conception of Innateness." *Philosophical Transactions of the Royal Society* B 366 (1563): 444–54.

McDonell, G. J. "Disciplines as Cultures: Towards Reflection and Understanding Transdisciplinarity." In *Recreating Integrated Knowledge*, ed. M. A. Somerville and D. Rapport, 25–38. Oxford: EOLSS.

Nagel, T. 2012. *Mind and Cosmos: Why the Materialist Neo-Darwinian Conception of Nature Is Almost Certainly False*. Oxford: Oxford University Press.

Niiniluoto, I. 1999. *Critical Scientific Realism*. Oxford: Oxford University Press.

Pinker, S. 2002. *The Blank Slate: The Modern Denial of Human Nature.* New York: Viking.

Pojman, L. 2005. *Who Are We: Theories of Human Nature.* Oxford: Oxford University Press.

Prinz, J. 2014. *Beyond Human Nature: How Culture and Experience Shape the Human Mind.* New York: W. W. Norton.

Psillos, S. 1999. *Scientific Realism: How Science Tracks Truth.* New York: Routledge.

Richardson, R. C. 2007. *Evolutionary Psychology as Maladapted Psychology.* Cambridge, MA: MIT Press.

Segerstråle, U. 2002. *Defenders of the Truth: The Battle for Science in the Sociobiology Debate and Beyond.* New York: Oxford University Press.

Smith, C. 2010. *What Is a Person?* Chicago: University of Chicago Press.

van Huyssteen, J. W. 1998. *Duet or Duel? Theology and Science in a Postmodern World.* Harrisburg, PA: Trinity Press International.

———. 1999. *The Shaping of Rationality: Toward Interdisciplinarity in Theology and Science.* Grand Rapids, MI: Eerdmans.

———. 2006. *Alone in the World? Human Uniqueness in Science.* Grand Rapids, MI: Eerdmans.

CHAPTER ONE

OFF HUMAN NATURE

Jonathan Marks

What I would like to do is articulate an anthropological position on human nature that is not official—there's no statement on human nature by the American Anthropological Association—but that I think is the most consistent with modern understandings derived from contemporary anthropology. I begin with a proposition about our place in nature—*nature*, of course, being a word that does a lot of work here—in the post-Darwinian world, namely, that we are biocultural ex-apes.

Ex-apes? Second part first. What do I mean by ex-apes? Evolution is descent with modification, and we are descended from apes, yet modified from them. Now, of course, it is not that difficult to find popular or even scholarly literature that will tell you that we are apes, on the grounds that our ancestors were apes. But that is not descent with modification; that is descent *without* modification. That is just descent.

To understand evolution means understanding that descent and modification are in constant tension with one another, and to reduce evolution to one or the other is a mistake. That's why when the evolutionary biologist George Gaylord Simpson said at midcentury, "It is not a fact that man is an ape, extra tricks or no,"[1] what he means is that there is

an important distinction to be made between our ancestry and our identity. What we are is not the same thing that our ancestors were. Our remote ancestors were fish, and although there are interesting things to be
learned about the human condition through an appreciation of our fish
ancestry, it really does not make any more sense to say that we are modified fish than it makes to call a book a modified tree. That emphasizes
the descent at the expense of the modification, and serves a purpose that
is rhetorical, not empirical. Moreover if science says that you are an ape
because your ancestors were apes, then what does science say about you
with the knowledge that your ancestors were slaves? Presumably, rather,
science would say in both cases that the identity of ancestors may at best
constrain but certainly does not determine the identity of descendants.

So that is what we are: ex-apes. Our ape ancestry can be seen if you
look hard for it. We share many of our anatomical features with apes.
A rotating shoulder, strongly flexed digits, and a short, stiff spinal col-
umn. Monkeys, by contrast, like many other groups of mammals, have
long, flexible spines, ending in a tail; extension of the wrist and fingers,
because they tend to walk on their hands rather than hang from trees; and
more limited movement at the shoulder.

And what do I mean by biocultural? About two and a half million
years ago, our ancestors started creating sharp surfaces, by deliberately
banging rocks together, and making not so much the first tools but the
first tools that are recognizable as tools in the archaeological record. So
we have been coevolving with, and adapting to, technology for millions
of years. And this shows up not only in the evolution of the tools themselves but in the evolution of our bodies as well. Chimpanzees do not
make nice things, because, compared to us, they have small, weak brains
and small, weak thumbs. Probably the only test of physical strength at
which you could beat a chimpanzee is a children's thumb-wrestling con-
test. In other words, (physical) dexterity coevolved with (cognitive) mentality and (inorganic) technology. That's a lot of coevolution.

And that is why we recognize that human evolution over the past
couple of million years reflects a change from the domain of biological
evolution to biocultural evolution, as our ancestors increasingly came to
rely on innovations in communication, technology, and society in order
to adapt and survive. And by about one hundred thousand years ago, our
ancestors were surviving not so much by what was in their heads as what

was *between* their heads, that is to say, relationships and historical traditions that transcend the individual organism and are in a classic sense, then, superorganic. And of course today we do the great bulk of our adapting culturally. Alleles for good eyesight and disease resistance tweak the gene pool, but glasses and antibiotics have had a much greater effect in solving those problems.

So as scientists interested in the human condition, so to speak, we focus on culture, because it embraces the historical process by which we principally adapt, and which gives us our orientation in a complex cognitive and social world. This is not to say that biology is not there in the modern world, but if you want to understand sugar consumption, it is insufficient to talk about our primate heritage. Our sugar consumption is the result of history, of political economy; like our meat consumption and beer consumption. They are all good in a biological sense, but a liter of Dr. Pepper is doing a lot more than satisfying a biological urge. And to understand it, you need to understand the relevant history, labor, pricing structure, availability, and so on. The biology, the nature, does not really tell you what you want to know in the modern world, even though it is there.

So if we are interested in explaining the major features of human behavior—why some people think it is OK to eat a dog and other people think it is OK to eat a horse, and still others think it is OK to eat a whale; why some people wear jeans while others wear saris—well, those are group-level differences, in which nature is a constant and culture is a variable. I mean, those populations are genetically slightly different but not in any sense that is explanatory given the question at hand. Consequently it is easy to juxtapose nature to culture and just work with the differences that are the products of history, not of microevolution.

Now I have just summarized the last century of anthropology for you in a couple of pages. The major patterns of variation in human behavior are found at the boundaries of human group identities, and they are the products of historical, rather than microevolutionary, forces. This is not to say that behavioral genetics is without value; it simply accounts for a trivial portion of the variation in human behavior. One person may be sadder and another happier for genetic reasons, but if those genes are patterned like the other genes we know of, then both forms are present in most populations. But if you want to know why one person is sad in

Chinese and another in French, genetics is not the place to look. Certainly the French depressive and Chinese depressive have far less in common in their lives than the French depressive and nondepressive or the Chinese depressive and nondepressive do.

Broadly speaking, we call that historical trajectory that structures lives and gene pools cultural, and acknowledge that it is a product of history, not biology. But might there be elementary structures of the human mind, upon which culture is simply inscribed? Perhaps it is human nature to think in terms of binary oppositions, as Claude Lévi-Strauss suggested: man-woman, day-night, hot-cold, dry-wet, human-animal, raw-cooked, nature-culture. Maybe that is a hard-wired aspect of human nature. On the other hand, archaeologists tend to think in trinary terms: Upper, Middle, Lower Paleolithic; Pre-Classic, Classic, Post-Classic; Stone, Bronze, Iron Age; Early, Middle, Late Pleistocene. But that would go against human nature—which would have to imply that either archaeologists are not human or they are unnatural. Surely there are a lot of people who think culturally in dichotomous fashions and there are many who do not. If that means that it is human nature to think dichotomously except when we do not, then it is not a very useful statement about humans, or biology. It might be a good description about certain systems of human thought, but of course we have no evidence at all that there is any innateness to it; this becomes a discussion of human nature without nature.

Culture in Us

So rather than bracket nature and set it off from culture, we tend instead to see how culture helps to construct nature. Culture is part of us in three ways, of which I've already mentioned the first, as an ultimate cause of the human condition, something that constitutes the environmental niche to which we have been adapting: the social, technological, physical, and cognitive features that our gene pools have been adapting to over a few millions of years. Second, it is a proximate cause of the human condition, that is to say, as the aspects of the environment that we experience as individual organisms—not simply the communicative and cognitive environments we are born into but also ecological stressors, which might reflect life at high altitude, or a high-fat diet, or the repetitive motions associated with a lifetime of manual labor, or simply growing up under a regime of social prejudice, which is known as embodiment in modern racial theory.

And third, culture permeates the entire prospect of seeking and claiming to find human nature in the first place. In 1972, *Pioneer* 10 was launched, carrying a picture of a man and a woman out beyond the solar system (fig. 1). The first question you may be asking after seeing it is, why did NASA use your tax money to send a picture of naked people into outer space? Obviously, their aim was to present a depiction of the senders in the event that the probe was intercepted by aliens. Equally obviously, the people depicted are not accurate representations of the people who actually sent the spacecraft but are instead representations of what those people wished they looked like. Further, it is the group that the senders intended to convey their own membership in—not the nation, not the ecosystem, not the neighborhood, but the species. That is why they chose a youngish, well-built male and female. What sense that might make to alien minds is a good question, as is the question of whether the male is issuing a generalized greeting of some sort ("Hi! Welcome to the galaxy!") or a warning ("Halt! This is a private nudist colony!").

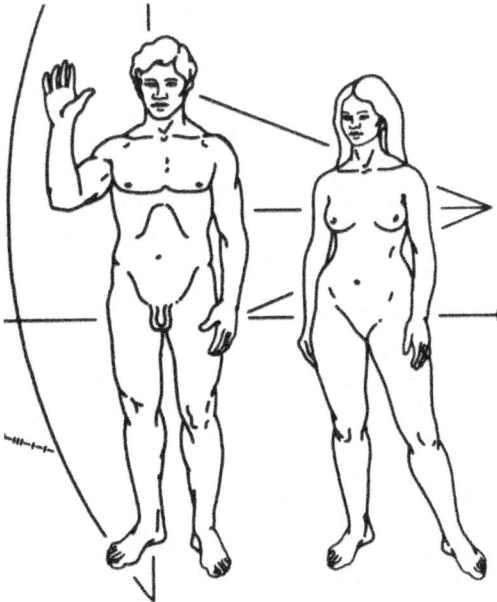

Figure 1. "Pioneer plaque" by Vectors by Oona Räisänen (Mysid); designed by Carl Sagan and Frank Drake; artwork by Linda Salzman Sagan. Vectorized in CorelDRAW from NASA image GPN-2000-001623. Licensed under Public Domain via Commons, https://commons.wikimedia.org/wiki/File:Pioneer_plaque.svg#/media/File:Pioneer_plaque.svg.

The larger point, though, is why they are naked and depilated. After all, when the aliens trace the probe back to Earth (a map is conveniently featured on another part of the plaque), would they not be surprised to find the humans clothed? Would they recognize us, if all they can match to the illustration is our face and hands? Or would they be angry at us for lying about what we look like and proceed to annihilate us on that basis? Why choose to send space aliens an image of what we look like and then show us differently from what we actually usually look like? The answer is that the astronomers intended to convey a "natural" image of our species, one stripping humans of their culture and depicting what they called in the French Enlightenment Man in a state of pure nature. And yet not only is that a lie, for that is not what the aliens will see when they land among us; but it is also a lie about a lie, for in imagining itself to be free of cultural information, the plaque nevertheless conveys cultural information. Certainly the haircuts and lack of body hair are cultural; as are the gendered postures and gazes, with only the man looking you straight in the eye (to a baboon, that would be a threat gesture; let us hope the aliens don't see it that way).

The point, then, is that culture is not to be scraped away, like the icing on a cake, but is rather an intrinsic part of the human condition, like the eggs in the cake. Not only is culture a part of us, but it is a part of how we see ourselves when we try to imagine ourselves without culture. It also means that opposing nature to culture is a falsehood, and thus to regard modern anthropologists as "idiot Lockean environmentalists" in order to make the case for human nature serves a rhetorical purpose but is without scientific merit or an empirical basis. The reason, then, that you find a lack of enthusiasm for a concept of human nature in anthropology is that "human nature" implies a separation from and opposition to culture, which was cutting edge in the nineteenth century but which was empirically seen to have outlived its usefulness over the course of the twentieth.

No Nature without Culture

Regardless, though, of the lack of formalization of a concept of culture prior to the nineteenth century, the idea that there is a discernible human

nature—perhaps rooted in genetics, rational thought, or social politics—
that corresponds to the dog's nature to bark, the cat's nature to purr, or
the pigeon's nature to nest and coo can be found among the ancient
Greeks. Of course, we realize that humans are different from other species, for example, by walking and talking. The question is how valuable a
pre-Aristotelian concept may be in a post-Darwinian world.

So when Mel Konner from Emory University asks rhetorically what
it could possibly mean to deny human nature,[2] in a journal called *Nature*,
I think there is a straightforward answer. If the assumption is that there
is an important or interesting part of the human condition that is brack-
etable and studyable independently of the effects of the cultural environment and social interactions with conspecifics, that assumption does not
seem to hold for much of anything. One of the most striking aspects
about human evolution is the change in life history that becomes obvious
when you compare humans to apes.

There has been a lot written lately on the evolution of menopause.
Where an ape breeds essentially until she dies, a human does not, and actually still has a few good decades left after the onset of infertility. The
value of this situation might be to assist her own daughter in parturition
and child-rearing, since those are a lot more difficult in humans than in
apes: the human infant's head is bigger than the ape infant's head, and
consequently where an ape mother can generally give birth by squatting
silently alone, a human generally has someone else around. Birthing is far
more social in humans than in our close relatives. Moreover, growth and
development are retarded in humans, so that where a chimpanzee is getting its wisdom teeth at age eleven, a human will wait another decade. So
not only are humans older for longer than apes are, but humans are also
younger for longer than apes are. Why? Because we survive by learning,
and learning takes time.

But one of the problems with the concept of human nature, imagining the isolation of those features that are innate from those that are
learned, is that those are not antonyms in human biology. The most un-
controversial aspects of our nature, walking and talking, are both innate
and learned. Not only are they learned in a generic sense, so that we can
communicate and locomote, but they are also learned in very specific cultural senses. We not only learn how to move, but we learn how to move
properly or appropriately. Ugandans and Japanese do walk differently. We

learn both language and a language, that is to say, the human form of communication and the locally meaningful and specific human form of communication. So it is not really that we are biologically programmed to walk and talk but that we are biologically programmed to learn to walk and talk correctly. By "correctly," what I mean is that there are a lot of ways of doing it but only one way of doing it right—so that the people around you can make sense of you, and you do not look or sound too weird.

The point is that we cannot juxtapose what we are biologically adapted to do against what we learn to do, because we are biologically adapted to survive by learning things, including most fundamentally how to locomote and communicate. But this fact raises an interesting question: If it is human nature to walk and talk, then you have defined two-year-olds out of the category "human." And based upon what standards of linguistic competence you adopt, you may be defining sixteen-year-olds out of our species as well. One possibility is to be a biological reductionist and define anything with human cells as human. Another, as some anthropologists have suggested, is to think about a human becoming, rather than idealizing a thirty-year-old hermit who is fully physically mature and yet devoid of any social experiences, as the proper referent of a human being and everyone else as an approximation of it. The essence of being human is to become, not simply to be.

Indeed, one of the important things that we become is social. We learn who is who and how to interact with them, and how not to interact with them. Although apes have kinship and humans have kinship, human kinship is governed by obligations and expectations, rooted in things like marriage and paternity, which do not exist in apes. Apes have mating, of course, and the difference between marriage and mating is that marriage produces in-laws, which apes do not have. The other thing about human kinship is that it is a calculating system. The knowledge that someone is from the clan or tribe of so-and-so, or descended from your great-grandmother so-and-so, or so-and-so's sibling or spouse, means nothing to apes but may specify quite precisely how you can interact with another human whom you may have never met; but knowing this about them enables you to situate them in an essentially formless social universe.

There are of course many ways of doing so. All societies have to juggle several variables in making sense of relatedness: marriage, generation, descent, sex, mother's or father's side of the family, and none of them maps perfectly well onto nature. For example, your father's brother and your mother's sister's husband are both your uncle, although only the first is a blood relative. This was one of the first big discoveries of anthropology—that for something as obvious and natural as relatedness, there was an amazing array of variation, none perfectly natural, but all providing rules for very precise social navigation in their respective cultures.

Are all variations possible or equally likely, perhaps constrained by the workings of the human mind? What does it mean about human nature that, say, 70 percent of the known cultures of the world can be coded as "polygamous"? Again, this is probably a poorly framed question, because it presupposes the reduction of systems of relations among people to the properties of the people themselves. But usually those relations are the products of history, not neurobiology. Consider, for example, the fact that in the Old Testament polygamy is pretty much taken for granted while in the New Testament monogamy is the relationship taken for granted. The difference seems to lie with those at least nominally monogamous Romans and the influence they cast in the region between the Testaments. The point is that, whatever the explanation, it is in the realm of social and political history, not individual brains.

The Biopolitics of Human Nature

Now, the biggest issues for me in accepting claims about human nature that often come from naive—that may be a euphemism—biologists involve normative behaviors. In his 1992 pop science best seller the biologist Jared Diamond explains that there is a relationship between sexual dimorphism (how different males and females look) and mating system in primates. At one end you have the highly polygynous baboons, where males are twice the size of females, and at the other end you have the monogamous gibbons, where males and females are the same size. On this scale, where do humans fit? Diamond explains that a "zoologist from Outer Space . . . would instantly guess that we belonged to a mildly polygynous species."[3]

Let us pause right there. Invoking fictitious extraterrestrial biologists is a problematic enough argument, but there is a normative statement here that needs to be rendered explicit. To say that we are naturally mildly polygynous is to say that if the biologist's wife comes home and finds him in bed with a female graduate student, he has an out: he was just expressing his mildly polygynous nature. But if he comes home to find his wife in bed with another man, well, excuse me, but we are *not* naturally polyandrous. That would be a crime against nature.

So the claim that our average body size says something about our basic inner natural social forms is not a value-neutral claim but a biopolitical claim. It is not strictly speaking false, but it is incomplete. After all, it is not just body size dimorphism that maps onto mating system. So does canine tooth size dimorphism. Those polygynous male baboons also have canine teeth twice the size of their female counterparts, while those monogamous male gibbons have canine teeth the same size as their female counterparts. And where do humans fit on this scale? Well, take any two people, and you will see that their canine teeth are about the same size.

In addition, human sexual dimorphism takes altogether different forms than you find in other primate species. Beards, breasts, body hair, body composition, these all help you tell a male human from a female human, yet have no homologous counterpart in the apes.

So where does that leave us in terms of the natural mating system that started this discussion? If you look only at body size dimorphism, you see us like polygynous baboons, as Diamond did. If you look at canine tooth dimorphism, you would instead see us like monogamous gibbons. If you look at body composition, you would see us as not even comparable to the other primate species.

I do not think that it is too radical a proposition to suggest that all this adds up to zero. We are capable of many diverse mating systems, and ethnographically that is exactly what we find: many diverse mating systems.

Even talking about human mating systems is a sleight of hand that makes us seem more comparable to other primate species. Other primates mate, humans marry; and obviously mating and marriage do not necessarily map onto one another particularly well in humans. What is interesting about marriage is that it produces social relations that other primates lack, namely, in-laws. That is why marriage is generally—ethnographically and historically—not so much a union of individuals as a union of families. And that is also why it is difficult to talk about the evolution of mate choice in humans if it is not clear that humans have evolved to choose their own mates, as opposed to having their mates chosen for them. Nevertheless, there is a rather extensive literature asserting that we have evolved a human nature that disposes men to be maximally attracted to women with a waist-to-hip ratio of .67, which just happens to be the last two elements of the classic 36-24-36 measurements of Hollywood starlets.

We are also, in this literature, attracted to people with symmetrical features, with wealth and power, and who are good kissers. Apparently our ancestors were drawn in many different evolutionary directions simultaneously back on the savanna, and none of them seems to have involved the reality of actually finding a partner that your parents would like.

Human Nature and Sex

As long as we are talking about sex, what about homosexuality? There is an extensive literature in biology purporting to try to explain it, from the standpoint that it is on the surface maladaptive, since someone gay might likely have a reproductive output lower on the average than someone straight. Now the problem here is the assumption that homosexuality is a genetic condition that requires some sort of evolutionary explanation, like sickle cell anemia. There are a number of assumptions here that can be interrogated, not only its genetic basis and pathological nature, especially from the standpoint of the evolutionary role of human sexuality. It assumes that the only natural function of human sexuality is reproduction.

Once again, this is biopolitics, so it is important to look carefully at the assumptions of human nature. Is the function of sexuality reproduction? Well, of course it is, especially if you are a lemur. Lemurs are only sexually active for a couple of weeks in the year, and there is no doubt that it is for reproduction. Other primates—monkeys, apes—are sexually active principally in reproductive contexts but in other contexts as well. The most extreme of these is the bonobo, in which sexuality is instrumental in creating social bonds specifically between females. Yet even they do not come close to approaching the human condition, in which the most basic aspect of human sexuality is its disjunction from the purely reproductive function it serves in lemurs. You can see this in many ways, of which the most obvious is the fact that in other primate species the males know just when the females are fertile. Their behaviors change, and sometimes they undergo physical changes as well. Compared to other primates, humans are quite extraordinary in the extent to which the female's reproductive status is concealed, both to her and to the male. If that were not true, then the rhythm method of birth control would work, and it does not.

Consider another feature of human evolution: breasts. An ape only has breasts when she is lactating, when they are filled with milk, and that is a strong signal to anyone else around that she is not fertile. In human adult females, breasts are filled with fat, whether or not she is fertile. So the cue of female fertility that an ape has is simply not available to the human. Human sexuality is characterized by far more extensive sexuality outside the context of reproduction. I am not trying to invoke the naturalistic fallacy here and tell you that you should engage in nonreproductive sex, although to be honest I should divulge that all the sex I have is with my wife, who is postmenopausal, and I hope you will not think less of me for being sexually active with her. The point is that descriptively, over the span of the human species and its sexual acts, a relatively small percentage actually involves the conjunction of the penis and the vagina of a fertile man and woman. Human sexuality is statistically far more about nonreproductive sex than about reproductive sex.

In this light, then, maybe homosexuality is not a genetically based pathology that requires an explanation as a maladaptive deviation from heteronormativity but rather, simply, another way of being nonreproductively sexual, which is quintessentially what humans have evolved to be.

Claims about Human Nature Have Ethical and Political Implications

Now you may be saying to yourself at this point, hey, that is political talk. Yes, that is exactly the point. To talk about human nature is to talk biopolitics. When you say that it is human nature for men to cheat on their wives but not for women to cheat on their husbands, and I say it is not, we are having a political conversation about gender roles in society. We can bring various arguments to bear on that question, some of them biological, but the point is that not only is it a political conversation, but it is a political statement to deny that it is political. This is of course exactly what the biologist is doing, as he claims to be just trafficking in value-neutral facts but presenting them selectively to make his point about the sexual double standard as a fact of human nature.

Some years ago, the economist Larry Summers was the president of Harvard and addressed the question why there were so few women on the senior science faculty at institutions like his own. His answer was that perhaps they lacked the intrinsic aptitude at the high end, which set off a bit of a controversy that eventually ended with his being replaced as president of Harvard by a woman. He and his supporters maintained, and probably still do, that he was a victim of politically correct antiscientific persecution. I mean, maybe it is true; maybe there are occult mental forces found more commonly in men that dispose them toward tenure in science. But how would we know, unless we examine the institutionalized practice of science at Harvard and elsewhere? Summers made that a non-question. By invoking imaginary genetic limitations, Summers was suggesting that the problem was not science's, or Harvard's, but women's. So if the phenomenon is simply a part of the biology of *Homo sapiens*, then there is really no problem of institutional discrimination worth worrying about or examining. Once again, we see a highly political dissimulation masquerading as a scientific fact of human nature. It has been wrong every other time it has been invoked, so the chances that it is right this time are probably pretty small.

Conclusion

To be human is to be cultural. Anthropologists are very engaged with their own history, to a much greater extent than other sciences are. The reason for this is that we have made all kinds of mistakes in the past; we do not want to commit them again. Science is not about repeating the mistakes of the past; it is about making new and creative mistakes. Whether it is the idea that human species comes naturally partitioned into a small number of flavors, or that brain size determines intelligence, or that we are simply to be understood as bipedal chimpanzees or hairy dolphins, or that our choices in mates were formed and passed down largely intact into the modern age because beautiful people outbred ugly people on the Pleistocene savanna, we have a good idea of what is false. It is of course harder to know what is true. In the present context, then, any discussion of human nature that segregates ostensibly innate (and often imaginary) genetic urges from the historical circumstances in which they become manifest has never been successful yet, and is simply unlikely to be so in the future.

Notes

1. George Gaylord Simpson, *The Meaning of Evolution* (New Haven, CT: Yale University Press, 1949), 283.

2. Mel Konner, "Seeking Universals," *Nature* 415 (2002): 121.

3. Jared Diamond, *The Third Chimpanzee* (New York: HarperCollins, 1992), 71.

RESPONSE I

On Your Marks . . . Get Set,
We're Off Human Nature

JAMES M. CALCAGNO

We are biocultural ex-apes.

—Jonathan Marks

Natural science will in time incorporate into itself the science of
man, just as the science of man will incorporate into itself natural
science: there will be one science.

—Karl Marx (1844)

To talk about human nature is to talk about biopolitics.

—Jonathan Marks

Politics is the art of looking for trouble, finding it everywhere,
diagnosing it incorrectly, and applying the wrong remedies.

—Groucho Marx (supposedly)

I have long admired the extraordinary ability of Jonathan Marks to convey ideas in ways that can turn an initially astonishing comment into
something so logical and unsurprising. Although I have given numerous

classroom lectures on some of the key issues covered in his essay, his comments always seem fresh, revealing, convincing, and illuminating. He is a tough act, and a pleasure, to follow.

Marks's expertise in anthropology and genetics has led him to two major conclusions: humans are the products of biocultural evolution, and we are biopolitical beings. Interestingly, although Karl Marx knew nothing about modern human genetics, he arrived at a similar perspective. Unlike Marks, Marx seemed convinced that humans had an essence, not surprising given the dominance of essentialism in his time. However, that essence involved our ability to make and shape our own nature (Marx 1844), which makes him sound more like a modern proponent of niche construction theory than an essentialist. As a result, like Marks, Marx appears more knowledgeable about human biology than many geneticists today. In addition, I would contend that Marx's prediction that "there will be one science" of humankind is coming true in the form of biocultural perspectives on human natures.

Marks's approach to his subject matter characteristically combines great scholarship with biting wit. In his essay his wit bites the big one, the heavily used and abused topic of human nature. In the last epigraph above, I ask the reader to substitute the phrase "The study of human nature" in place of "Politics" to see if the original sentiment changes much if at all. Human nature has been, and continues to be, a biopolitical minefield. Ironically, even the quoted quip is commonly misattributed to Groucho Marx, which to me serves as an added reminder that when enough people think something is true, or advance the same erroneous information repeatedly, it becomes common knowledge. Regrettably, however, I hasten to add that false ideologies of human nature continue to affect people in ways that are more far tragic than comic.

In any case, a biocultural approach to understanding human nature had led us to view human "natures" in the most flexible terms. This view in many ways often resembles a "blank slate," with "nurture" decisively prevailing over "nature" more than this biological anthropologist would have once imagined. Although I have always admired and promoted biocultural perspectives, I have never admired or embraced the blank slate concept. Taken to its extreme, all species are blank slates, which cannot possibly be true. In my house everyone talks to the dog, yet it still doesn't speak a word of English beyond "rough." And our hamster has never once shown the slightest concern about what is on my mind, not even pretending to have an interest in how my day went, like the rest of my family.

Given that no other animal is or can be a blank slate, I cringe at any thought of isolating humans to a special status that separates our species from other organisms, as if exempt from the evolutionary forces acting upon all other life-forms. As a biological anthropologist, I also shudder at the notion that biology may be irrelevant to what makes us human, as many of my more science-anxious colleagues have contended for so long. Yet culture seems to bring us as close to a blank slate as one can imagine. What is an evolutionary anthropologist to think?

Although I could answer that question by saying, pay close attention to what Marks has written on the subject here and elsewhere, I want to take advantage of my opportunity to comment further. First and foremost, I would argue that a central part of our human nature, what makes us human, is how we learn about the world around us, including, most notably, other humans. How we learn, though comparable at some levels to other species, is unique and done in unprecedented ways on this planet. Our way of learning to navigate through our world has led to such remarkable behavioral diversity that it is little wonder the concept of a blank slate has been so appealing to so many. At the same time, I contend that our evolved learning mechanisms have ironically predisposed us to make great mistakes in understanding our own (and other) species. Thus, in the brief space available, I attempt to explain why the way we learn is so important, why a biocultural approach that cuts across all disciplines is providing hope for a better understanding of what makes us human, and why it is important to correct mistakes in our understanding of ourselves.

Learning

Although I agree with Marks that we are not programmed to think and learn in terms of dichotomies, one must admit some irony there after discussing whether we are apes or ex-apes, the long history of juxtaposing nature to nurture, all in the context of who is right or wrong. Indeed, we commonly go beyond paired comparisons, as in a divine Holy Trinity, and the amazing biblical prophecy that came true on the Notre Dame campus, the arrival of the Four Horsemen. However, whether we deal in dichotomies, trinities, centuries, entities, or identities, we are predisposed to categorize and to create clear boundaries often where no boundaries exist.

As one example we can take a continuous color spectrum and divide it into seven obvious colors, which we may remember by turning their first letters into an acronym; "Roy G Biv" attests that even a dreadfully painful acronym works better for minds searching for patterns than nothing at all. Science can then verify that these colors do indeed correlate with wavelengths of light, or so it seems. Yet if the range of green wavelengths is 480–560 nanometers (nm) and yellow ranges from 560 to 595 nm, what exactly is the color of 560 nm, and when is the last time you could not tell the difference between yellow and green? Most likely it was when you saw something around 560 nm, and you are unlikely to remember when that was, but you can recall many times when you clearly saw yellow or green because of your cultural categories.

On the upside, categories often tell us a great deal about how to operate within the world around us: that's land, drive on it; that's Lake Michigan, don't drive on it. On the downside, we often create categories that do not exist and from them extract totally mistaken notions, with appalling consequences. I would argue that our biocultural evolutionary past turned us into innate essentialists, looking for the categories and the essence of colors, species, races, kin, and everything around us. Evolutionarily we have done this in a "fast and frugal" way (Richerson and Boyd 2005), because it is too costly for past hominins and living humans to try to learn everything on one's own. Thus quick judgments are made: this person is kin so I trust, that person looks and dresses different from kin so I'm more wary. Of course all of this profiling can be overturned by positive or negative experiences with those you think you can or cannot trust, if given the opportunities. I am not suggesting, then, that we are biologically destined to think in such ways, but we are predisposed to learn in essentialist patterns, making essentialism past and present predictable, understandable, and easily absorbed by the masses. As a result, although more modern populational approaches today stressing variation over typology can be far more reasonable, they are likely to be far less intuitive or obvious.

Learning about Ourselves

Given that populational thinking has characterized all of my courses over the past thirty years, it is hard to imagine a single lecture with the term *variation* not being used in some context, whether in regard to fossil evidence or primate behavior, human genetics, cultural diversity, and so on. In addition, throughout this time I feel I have produced a constant drumbeat of breaking down essentialist categories and barriers: between races, fossil species, and humans and nonhumans. With regard to the latter, my emphasis has always been on the similarities we share with nonhuman primates, often with eye-opening effects on many students. Thus it feels odd to focus on human discontinuity with other species from an evolutionary perspective. Yet even those who embrace the concept of cultural primatology must surely recognize, like Jonathan Marks, that "ape culture" is not the same as "ex-ape bioculture," just as morphologists do not equate occasional bipedalism in apes with habitual bipedalism in humans. Chimpanzees may fashion tools to fish for termites, but they cannot comprehend why anyone would dress like African explorers to publish papers on why another species would make such tools.

Given this discontinuity, has biology become trivial and has culture turned us into blank slates? Our desire for either a yes or no response may again illustrate how we want to learn, using clear-cut, mutually exclusive categories of nature or nurture. Regardless, genes are certainly trivial when looking at humans the old-fashioned way: grouped into genetic races, or employed to suggest why one culture is behaviorally predisposed to say yay or neigh to horse meat. Large genetic subspecies do not exist in humans, and genes build proteins, not cultural nuances. However, biology is still highly relevant in important ways, which may be missed in much of today's anthropology, and I will provide two examples.

Biocultural Approaches as the "One Science of Humankind"

In 2011, the psychologist Nalini Ambady wrote a review of what she labeled the "emerging field of cultural neuroscience," a phrase that might initially send chills up the spine of many anthropologists, fearing a new version of biological determinism. However, in her next sentence she wrote, "Increasing evidence of the brain's plasticity, the evolutionary basis of cognition, and the co-evolution of culture and the brain makes clear that cultural and neural processes are interwoven" (2011, 53). At the end of her review she noted both the challenges and the need to study diverse cultures and to "train young researchers in both cultural psychology and related fields (e.g., anthropology) as well in the theories and methods of neuroscience" (80).

In 2013, George Slavich and Steven Cole wrote an essay titled "The Emerging Field of Human Social Genomics," which again might lead many anthropologists to immediately think of eugenics. Yet the authors' main premise reads, "our genome appears to encode a variety of 'potential biological selves,' and which 'biological self' gets realized depends on the social conditions we experience over the life course" (2013, 331). They then draw a direct contrast to a nature versus nurture mentality, for the following reason: "Studies that elucidate how the external social environment gets translated into the internal biological environment of disease pathogenesis may lead to new methods for preventing disorders and promoting wellness, but much more research is needed to turn this promise into a reality" (343).

Here we see the promise of biology used not as a social weapon but to understand social cures. The biocultural perspective goes well beyond the subfields of anthropology while aiding and abetting anthropology's promise of a more holistic approach to the study of humanity (all of which leads to my final point).

Our Biocultural Past and Future

Although some might disagree about the exact wavelength range of indigo and whether it should be eliminated from the spectrum of continuous color variation, arbitrary categories of people being eliminated despite the spectrum of continuous human variation is a far different narrative. When diseases are investigated not by racial typology but by cultural circumstances, when more people understand that biology is not identical to genetics and that our nature is wholly dependent on our culture, different avenues of reasoning can open minds. The difficulty is that we are asking human minds predisposed to seek distinct categories and their inherent essences to not go with what seems so intuitive. Yet an education to override misconceptions and misunderstandings of human nature can be achieved, and with increasingly rapid results given the potential speed of communication.

For better and at times for worse, scientific perspectives of what makes us human carry great weight today, so all the more reason to advance a twenty-first-century biocultural view of humanity and to debunk some nineteenth-century ideas still offered today that may be more attractive because of their simplicity and inaccuracy. With some success, we do that in our classrooms, but not all students get it, and fewer transmit it. To increase the transmission of biocultural perspectives on human natures, conferences such as the one leading to this volume can be an excellent start. However, in my opinion, for such gatherings to become important conferences, ideas must go beyond the meeting to much larger audiences. The issues being dealt with here are more than just intellectually stimulating for a few. They are potentially liberating for many. I hope these efforts will not only lead to an improved understanding of humanity but also provide some contribution over time to improved interactions within humanity.

ACKNOWLEDGMENTS: Groucho Marx did famously say, "I don't want to belong to any club that will accept me as a member." In this case, I could not disagree more, as it was truly a privilege to be part of the conference and the resulting publication. I especially want to thank Agustín Fuentes and Aku Visala for the invitation and for making both the conference and this volume possible. I also owe special thanks to Jon Marks for providing a remarkable essay and for always making me think harder about what I thought I knew.

References

Ambady, N. 2011. "Culture, Brain, and Behavior." *Advances in Culture and Psychology* 2: 53–89.

Held, J. M. 2009. "Marx via Feuerbach: Species-Being Revisited." *Idealistic Studies* 39 (1–3): 137–48.

Marx, K. 1844. "Private Property and Communism." [Appendix to incomplete Second Manuscript of Economic and Philosophic Manuscripts.] Available at www.marxists.org/archive/marx/works/1884/manuscripts/comm.htm.

Richerson, P. J., and R. Boyd. 2005. *Not By Genes Alone: How Culture Transformed Human Evolution.* Chicago: University of Chicago Press.

Slavich, G. M., and S. W. Cole. 2013. "The Emerging Field of Human Social Genomics." *Clinical Psychological Science* 1 (3): 331–48.

RESPONSE II

Rethinking Human Nature

Comments on Jonathan Marks's Anti-Essentialism

PHILLIP R. SLOAN

Jonathan Marks has given us a provocative and controversial discussion of a concept that has played, and continues to play, a major role in post-seventeenth-century science, philosophy, politics, and theology. Generally, he is concerned with the question of whether it is meaningful to speak of "human nature," and if so, how it is to be conceptualized. His general conclusion is negative.

His approach as a biological anthropologist and human evolutionary biologist also defines the parameters of his discussion. In this case we are not considering issues of fundamental epistemology, or metaphysics, or a comprehensive philosophical anthropology that might draw on sources other than empirical science. Rather than raise criticisms from these alternative perspectives, I accept the framework offered and analyze the strengths and weaknesses of Marks's argument, taking into account its obvious restrictions.

Darwin Deleted

I preface my remarks with an observation. In his recent imaginative counterfactual history, *Darwin Deleted* (2013), the Darwinian historian Peter Bowler reflects on what the conversation about evolution and its extensions today would be if Darwin had simply fallen overboard from the HMS *Beagle* and drowned. His speculative but plausible conclusion is that in many ways we would still be theoretically in biology much where we are. Several lines of work in the mid-nineteenth century were converging on the thesis of descent with modification from common points of ancestry, although this might not have come together in a general theory of common descent until the 1870s; Alfred Russel Wallace, as is well known, had independently arrived at the principle of natural selection by the summer of 1858, much to the consternation of Charles Darwin, and his views would likely have affected the development of evolutionary theory, although somewhat later. Social Darwinism, already on the scene with Herbert Spencer, would likely have developed from more purely Spencerian roots.

Within this fictional scenario, we would predictably today be holding a somewhat different conversation about the place of human beings in an evolutionary world, and thinking a little differently about how this related to a conception of human nature. Where the great difference would lie between our world without Darwin and the one we actually inhabit is the interpretation of how physical continuities and even physical genesis are related to the human world—the domain of cultural, mental, and moral evolution. Wallace, for example, concluded in an important series of papers in the 1860s that natural selection could only refine the human frame and brain to a certain minimal improvement over primate ancestors. From that historical point, very different forces related to language and culture would have taken over. This would have given us a human evolutionary story generally cut off from natural selection in Darwin's sense.

But Darwin did not drown. Instead his *Descent of Man* of 1871, followed by the *Expression of the Emotions* in 1872, changed the parameters of this possible alternative dialogue in dramatic ways. All significant human properties, including not only skeletal and physiological features but also mental, moral, aesthetic, social, and even religious dimensions of humans, were to be viewed by Darwin from the standpoint of zoology and natural history and explained by their genesis from animal origins under the dual action of natural and sexual selection. There was no residual left for nonbiological factors to explain in terms other than those reducible to his two selective forces. Darwin's solution was ruthlessly reductive, if not in the modern sense of biophysical reduction. It was instead a reduction of human properties to their zoological roots in other forms of life (Sloan 2015).

Although the fate of Darwin's solution to the question of human evolution has been checkered and nonlinear, with the intervening era dominated by Franz Boas and his successors, as described by Robert and Linda Sussman in this volume, Darwin's perspective has reemerged in recent decades with a vengeance. Today we are confronted by the flood of works, both popular and professional, in sociobiology, evolutionary psychology, primatology, and behavioral genetics that take as a premise the Darwinian claim that the genetic continuity of life warrants genetic explanations of all significant human properties.

Ancestry Is Not Identity

Marks takes strong issue with this pervasive "biologizing" of the human lifeworld, and in this I would certainly agree with him. His argument is in many ways similar to that of Wallace earlier. Specifying historical continuity, typically meaning skeletal and general biological continuity, does not amount to an adequate explanation of what has happened in the last hundred thousand years of human history: "An important distinction is to be made between our ancestry and our identity," as he puts this. As Marks has argued in his book *What It Means to Be 98% Chimpanzee* (2002), the near-identity of human and chimpanzee DNA does not mean that we have explained something truly interesting about human beings when this elementary level of genetic connection is specified.

The period of primary interest to Marks—the last hundred thousand years of human history— is that in which we find the profound interaction of language, technology, symbols, self-reflection, and the development of complex forms of social interaction. These have created human history. And in Marks's view, all this *does* make a profound difference in deciding who we are as human beings, even if we may have historically descended from other primates.

The central question addressed by Marks concerns what the recognition of cultural evolution, with all this entails, means for a concept of human nature. The issue is, and has been, a classic boundaries debate. A long history has assumed that there is a given biologically defined universal human nature that can be discovered by stripping away culture—pure zoological humanity in the state of nature. To this is then added the impact of history, education, social custom, language, and reflection. Certainly this familiar contrast involves many questionable assumptions. Even Jean-Jacques Rousseau, in creating this story of the developmental history of humanity in his famous *Discours sur l'inégalité* of 1755, presented it as a counterfactual story, a way of analyzing out some of the issues of contemporary society through a historical story of its development. Since Darwin, this account has assumed greater force, now accepted as the dictate of modern biological and human science. The relative weight given the Wallace and Darwin options has in many ways defined the conflict of "biological" and "cultural" anthropology.

Marks resolves this conflict by arguing against the concept of human nature itself: "the reason . . . that you find a lack of enthusiasm for a concept of human nature in anthropology is that human nature implies a separation from and opposition to culture." In its place he argues for a concept of cultural becoming through learning and socialization: "The essence of being human is to become, not simply to be" (Marks this volume).

How Far Off Human Nature?

Much of Marks's argument here I can certainly agree with, and he has drawn our attention to what it is that makes us other than "naked apes." But what is the shape of this third option of cultural becoming that he seems to press us toward? Here I think we need some greater clarity of terms.

First, we need some clarity about the meaning of "human nature." This hoary concept, important for the tradition of Aristotelian, Stoic, and

Scholastic ethical and anthropological reflection, needs not be conceptualized simply in purely biological terms and has not been so conceived in its history, as seems to be presumed in Marks's discussion. For an Aristotelian, for example, such a human nature would imply an integral relation between the material conditions of life and the immaterial formal organizational aspects that together would constitute something like human nature. What I think is needed is not the abandonment of the concept but a deepening of what we understand it to mean.

The social and political events of the post-Enlightenment period have allowed us to see the outcome of the denial of some substantive and normative conception of human nature. We can view this as a historical slide from a traditional notion of natural law, to the assumption of a stable human nature, to the fragmentation of this concept in the early nineteenth century (Sloan 1999). The human species, as a result, was pluralized into different historical races, and even different species, standing in the relation of degeneration to one another (Sloan 2014; 1973). The concept of race was been "essentialized" to imply fixed and unalterable differences, as argued by Kant and Blumenbach (Bernasconi 2001; Sloan 2014). Scientific polygenecists in the nineteenth century could then expand on the thesis that there are different biological species of human beings, a claim that became widespread in the secular scientific literature of the time as a warrant for human slavery. Even Darwin, who, as Adrian Desmond and James Moore have argued, opposed polygenecism and was personally motivated in developing aspects of his theory by his distaste for human slavery (Desmond and Moore 2009), nonetheless formulated arguments in the *Descent of Man* on the evolution of human races that left the final conclusion to be drawn from his analysis highly ambiguous. Whether human races, evolving under the pressure of group selection, have attained the level of distinct species or are mere varieties within one species, is not clear from his arguments. The eugenics movement, drawing on theories of genetic determinism and racial ideology, drew conclusions from these premises that had horrendous consequences in the last century familiar to us all.

I point out these historical consequences that have in fact resulted from the denial of a substantive concept of human nature that, for all its deficiencies, marked one of the admirable developments of the Enlightenment on which democratic ideals were created. Simply abandoning this concept, succumbing either to complete social constructivism or to trans-humanism, has consequences that cannot be embraced without serious critical reflection.

Marks indeed is not advocating these extreme options, and mainly wishes to move out from under a flat, ahistorical view of human nature as some kind of crystalline biological given. I would agree with this intent, but I would suggest that Marks has not taken sufficiently into account what is implied by his comment that "the essence of being human is to become, not simply to be."

Becoming and Teleology

Can we give any further meaning as to what this "becoming" might be directed toward? For a neo-Aristotelian , it would have some meaning: it is to become fully human as an embodied being comprising both dynamic inner properties and a biological material being. Such a being seeks full integration of biological and physical properties with those of consciousness and ethical reflection. But this is on the assumption that there is a realization of some inner end or "nature" that constitutes the all-important inner being of the human, something that is not reducible to genes and physiology and yet retains a fundamental reality.

It is not clear, however, how Marks envisions this human becoming. Here he may be trapped by the anti-teleological conclusions that so often accompany a Darwinian perspective—a becoming that has no goal, one that simply involves a kind of divergent adaptive "speciation" in response to local conditions, even if this development is a cultural rather than simply biological phenomenon. But this is, in my view, the great problem that confronted, and still confronts, Darwin's evolutionary ethics and its contemporary successors: can we define some notion of the "good" that is more than adaptation to local conditions and survival advantage? This requires that the "nature" involved is something more than a collection of properties defined by external circumstances, whether biological or cultural. It requires some unification by an inner nature-with-purpose. This may include biology but biology as subordinated to an inherent teleological end.

The issue seems to be whether there can be, in spite of our acceptance of an evolutionary account of human origins, a definition *within a given time horizon* of a sufficiently stable conception of human nature that allows it to function for the wider purposes to which the concept has been applied in the past: as a ground for a concept of universal human rights; as a basis of ethical realism; as the warrant for some normative sense of human dignity that transcends specific cultures. This does not mean that it cannot be flexible enough over time to offer some framework for asserting the historical progress and change of human sensibility, or even a fundamental evolutionary change in a longer history of genus *Homo*. But it does mean that within a synchronic slice of history there is some useful reference to the concept of human nature.

To make this claim, we first must dispense with the tired refrain that evolutionary theory "destroyed essentialism." This mythic story cannot, I would argue, be maintained, and the issue has simply been badly formulated in the literature of evolutionary biology.[1] What is required is a more dynamic and multilevel concept of human essence or nature that would still be able to function meaningfully for wider philosophical purposes. It seems defensible to claim that "becoming" could include a potential drive of humanity toward some ideal goal that is more than adaptation to immediate circumstance, whether biological or social-cultural.

Perhaps this can best be stated only in terms of a regulative Idea in Kant's sense as it relates to our concrete, multicultural practical life—a demand of rational reflection that we cannot expect to instantiate by force or legislation but that nonetheless can direct us as a working teleological end of human achievement, supplying some way of sorting out *which* cultural developments help us achieve this end. Such ideals are embodied in documents like the United Nations Universal Declaration of Human Rights and in the social teachings of the Roman Catholic tradition. Without some concept of human nature to which such rights and ethical norms adhere, can any sense be made of such appeals? To answer this affirmatively requires neither a denial of culture nor a denial of evolutionary biology. But it does imply some realization of human transcendence over simple biological heritage, and also some realistic framework for normative value.

Jon Marks has done important and landmark work as a biological an-thropologist in addressing some of the biological reductionism that sur-rounds us today. In these remarks I hope I have been able to indicate as well some of the ways in which we need not abandon some important traditional concepts that can still play an important role in our larger philosophical, ethical, and theological discourse.

Note

1. This claim, first made by John Dewey in 1909, has been widespread in the philosophical literature since the 1965 article by David Hull (Hull 1965) and is commonly repeated as an axiomatic principle of neo-selectionist evolutionary biology. This claim has been attacked from different directions by a wide range of contemporary scholars, including Oderberg (2007), Winsor (2006), Wilkins (2009), and Richards (2010). I have reviewed this issue in Sloan 2013.

References

Bernasconi, R. 2001. "Who Invented the Concept of Race? Kant's Role in the Enlightenment Construction of Race." In *Race*, ed. R. Bernasconi, 11–36. Malden, MA: Blackwell.

Bowler, P. J. 2013. *Darwin Deleted: Imagining a World without Darwin*. Chicago: University of Chicago Press.

Desmond, A., and J. R. Moore. 2009. *Darwin's Sacred Cause: How Hatred of Slavery Shaped Darwin's Views on Human Evolution*. Boston: Houghton, Mifflin, Harcourt.

Hull, D. L. 1965. "The Effect of Essentialism on Taxonomy: Two Thousand Years of Stasis." *British Journal for the Philosophy of Science* 15: 314–26; 16: 1–18.

Marks, J. 2002. *What It Means to Be 98% Chimpanzee: Apes, People, and Their Genes*. Berkeley: University of California Press.

Oderberg, D. S. 2007. *Real Essentialism*. New York: Routledge.

Richards, R. A. 2010. *The Species Problem: A Philosophical Analysis*. Cambridge: Cambridge University Press.

Sloan, P. R. 1973. "The Idea of Racial Degeneracy in Buffon's *Histoire naturelle*." In "Racism in the Eighteenth Century," ed. H. E. Pagliaro, 293–321. *Stud-*

ies in Eighteenth-Century Culture 3. Cleveland, OH: Case Western University Press.

————. 1999. "From Natural Law to Evolutionary Ethics in Enlightenment French Natural History." In *Biology and the Foundations of Ethics*, ed. J. Maienschein and M. Ruse, 52–83. Cambridge: Cambridge University Press.

————. 2013. "The Species Problem and History." *Studies in History and Philosophy of Biological and Biomedical Sciences* 44: 237–41. Available at www .sciencedirect.com/science/journal/13698486/44/2.

————. 2014. "The Essence of Race: Kant and Late Enlightenment Reflections." *Studies in History and Philosophy of Biological and Biomedical Sciences* 47: 191–95. Available at www.sciencedirect.com/science/article/pii/S136984 8614000752.

————. 2015. "Questioning the Zoological Gaze: Darwinian Epistemology and Anthropology." In *Darwin in the Twenty-First Century: Nature, Humanity, God*, ed. P. R. Sloan, G. McKenny, and K. Eggleson, 232–66. Notre Dame, IN: University of Notre Dame Press.

Wilkins, J. S. 2009. *Species: A History of the Idea*. Berkeley: University of California Press.

Winsor, M. P. 2006. "The Creation of the Essentialism Story: An Exercise in Metahistory." *History and Philosophy of the Life Sciences* 28: 149–74. Available at www.hpls-szn.com.

RESPONSE III

Off Human Nature and On Human Culture

The Importance of the Concept of Culture to Science and Society

ROBERT SUSSMAN AND LINDA SUSSMAN

From Eugenics to Culture

In the early 1900s, the eugenics movement in the United States and Western Europe had divided the world into fit and unfit individuals. Western Europeans and those Europeans who had first migrated to the United States were the most fit. The unfit were peoples from other countries, as well as those with some kind of physical or behavioral disorder. The biologist Charles B. Davenport led the American eugenics movement, but Madison Grant, with his *Passing of the Great Race* (1916), defined its policies and goals. Basically, "fit" people were genetically superior to other people, and only Western Europeans were able to establish and sustain a civilized society. All other individuals were less fit genetically, and, by some eugenic measure, their populations should be limited or eliminated. Eugenicists believed that most behavioral features were biologically fixed and could not be changed by the environment (Sussman 2014).

However, in 1911, during the early part of the eugenics movement, Franz Boas wrote two books that challenged the current concepts of eugenics. In *Changes in the Bodily Form of Descendants of Immigrants* (1911a), he showed that the head shape of immigrants could change within one generation, head shape being one of the fixed bodily forms that "defined" people and "races." In *The Mind of Primitive Man* (1911b), Boas developed the anthropological concept of culture. In this book, using his and others' ethnographic research, he claimed that mental attitude was not determined by heredity, that any people or race could achieve civilization given proper environmental conditions, that there was more variation within races than between them, and that environment accounted for the differences between various peoples and so-called races. From the early 1900s to the 1940s, there was a major battle between the racist American Eugenics Society and Boasian anthropology. It was only between the 1940s and 1960s that the Boasian anthropological concept of culture began to become the common paradigm in science. This was an important period in the history of modern anthropology (Spiro 2009; Sussman 2014).

Culture, as defined by Boas, is a set of guidelines (both explicit and implicit) that individuals learn as members of a particular society and that tells them how to *view* the world, how to experience it emotionally, and how to *behave* in it. Through enculturation, an individual slowly acquires the cultural "lens" of that society. Without such a shared perception of the world, human society would be impossible (adapted from Helman 1994). Most of what people learn is filtered through that lens, and people brought up in the same culture share, more or less, a common perception. The differences among people are the product of different histories and not differences in their biology (Degler 1991). People within different cultures, ethnic groups, or, in some cases, different subcultures share the same worldviews, and these views may or may not be shared between groups. Cultural differences are extremely important. The concept of culture was a genuine seminal contribution (Deglar 1991). By the 1940s and 1950s, the concept of culture became a major scientific paradigm for anthropology and other social sciences.

Animal Culture?

Given the importance of cultural differences among people, what can we say about these differences among nonhuman animals? Do animals have culture? Corvids—crows and ravens—have been identified as having culture. Marzluff and Angell (2005) claim that these birds have a number of different behaviors indicating culture, the most persuasive being the propensity for New Caledonian crows (*Corvus moneduloides*) to use different types of extractive foraging tools depending on the geographic area in which they live. Here culture represents the ability to socially transmit a behavioral pattern from one locality to another. Male humpbacked whales (*Megaptera novaeangliae*) sing similar breeding songs on grounds thousands of kilometers apart, but the particular songs in different areas are different. Male bowhead whales (*Balaena mysticetus*) also have different song dialects in different breeding areas. Some bottlenose dolphins (*Tursiops* spp.) have been observed carrying sponges (Rendell and Whitehead 2001). These behaviors have been seen as indications of culture, since the information or behavior is acquired from conspecifics through some form of social learning (Boyd and Richerson 1996). Meerkats (*Suricata suricatta*) have been seen teaching their pups prey-handling skills by providing them with opportunities to interact with live prey. Thornton and McAuliffe (2006) claim that the provisioning behavior of meerkat helpers constitutes a form of "opportunity teaching" and that this perpetuates socially learned "cultural" skills.

The evidence for "culture" in nonhuman primates has a long history. This began when Japanese primatologists noted the acquisition and passing on of potato washing with river water and then dipping the clean potatoes into salty seawater among macaques (Kawamura 1959). Later the macaques were observed to throw wheat into the water and wait for it to float back up before picking it up and eating it free from dirt (Nakamichi et al. 1998). Since then, a number of similar observations have been made. Capuchins have adopted a number of local traditions, like eating some foods at some sites and not at others, sniffing each other's fingers and then inserting them into one another's noses, and even poking each other's eyeballs (Perry et al. 2003; Balter 2010). Santorelli, Schaffner, and

Aureli (2011) observed spider monkeys with twenty variant behaviors observed across a number of sites, each with differing degrees of prevalence among various community members. Soon after an outbreak of tuberculosis, a baboon group was left with males who were less aggressive and more social than average baboons, and the group doubled its previous female-to-male ratio. Aggression was less frequent and rates of affiliative behaviors, such as males and females grooming, soared. Even after twenty years, with newly immigrated males, the group's unique social milieu persisted. The group's low aggression/high affiliation society constitutes nothing less than, according to Sapolsky (2011), a multigenerational benign "culture." Van Schaik et al. (2003) list putative cultural variants at six orangutan sites containing twenty-four elements. Ten of these involve specialized feeding techniques (including tool use), and six are alternative forms of social signals (such as kiss-squeaks).

But it is the chimpanzee that has been given the most attention concerning nonhuman animal culture. Whiten and colleagues (1999) found thirty-nine different behavior patterns distributed among sites of the seven longest-term chimpanzee studies. These patterns present a systematic synthesis of information from an accumulated 151 years of chimpanzee observations. They include behaviors on tooth usage, grooming, and courtship. Such behaviors as pounding of food onto wood, pounding of food onto other hard surfaces, termite fishing using leaf, termite fishing using non-leaf, leaf clip, mouth-leaf clip, fingers-leaf clip, hand clasp, aimed throw, self tickle, stem pull, leaf-napkin, leaf-dab, branch din, and branch-slap were distributed differently in different chimpanzee groups (see Whiten et al. 1999 for a complete list of behaviors). Whiten et al. claim that in the biological sciences cultural transmission is recognized when intergenerational transmission of behavior may occur either genetically or through social learning. The processes of variation and selection shape biological evolution in the first phase and cultural evolution in the second. Using this definition, a cultural behavior is one that is transmitted repeatedly through social or observational learning to become a population-level characteristic. Again, using this definition, "cultural differences (often referred to as 'traditions' in ethology) are well established phenomena in the animal kingdom and are maintained through a variety of social mechanisms" (Whiten et al. 1999, 682).

For Whiten and colleagues each of these behaviors refers to a variation in one (or a few) single behavior(s). However, in chimpanzees population differences have indicated that multiple behavioral variants exist. Some have claimed that the differences in chimpanzee groups can be explained by genetic differences between subspecies. However, the results show that the behavioral patterns in many cases are the product of social learning and, therefore, are considered "cultural" behaviors (Lycett, Collard, and McGrew 2007). So for many primatologists, as de Waal writes,

> all in all, the evidence is overwhelming that chimpanzees have a remarkable ability to invent new customs and technologies, and that they pass these on socially rather than genetically. . . . [T]he 'culture' label befits any species, such as the chimpanzees, in which one community can readily be distinguished from another by its unique suite of behavioral characteristics. Biologically speaking, humans have never been alone—now the same can be said of culture. (De Waal 1999, 636)

Culture and Human Nature

However, is this what Boas meant and what many anthropologists and social scientists mean when they use the term *culture*? In our view, we are simply dealing with a complete lack of understanding, and this understanding is extremely meaningful when we try to understand human versus animal behavior. To us, this is similar to the problem one has when one says that chimpanzees and gorillas communicate with language. Yes, chimpanzees and gorillas do have some of the attributes that it takes to communicate using language. However, these are only a small part of the large number of characteristics needed for human language to exist (Hockett 1960; Hauser et al. 2002). No one can say that what chimps and gorillas do is language. For example, recurrence is one of the characteristics of what humans do that cannot be done by chimpanzees or gorillas. After the age of about three and a half, human children are able to count, and they understand the system of counting. In contrast, it takes an adult chimp thousands of hours to learn to count to nine. Each number takes the same amount of time to learn. If there is any lapse in time, the whole system of counting must begin again and a complete re-learning is required. At around three and a half, a human child catches on to 1, 2, 3 (and sometimes 4) and quickly acquires all the other numbers. The human child grasps the idea of an integer list. Chimpanzees never learn the open-ended generative properties of counting (Hauser et al. 2002). This is a necessary part of language, and recurrence seems necessary also in such tasks as math, human-generated kinship systems, and music.

Chimpanzees do not understand abstract thought and hierarchies of increasing complexity. In the past, language has always been referred to as a human phenomenon. If what chimpanzees and gorillas do is called language, then a new term is needed for the unique things that humans do. We would prefer to stick with tradition and call what humans do language. Let's find another term for what chimps and gorillas do. We might be able to learn something about what humans do by studying this behavior, but what chimpanzees and gorillas do is *not* language. What chimps and gorillas do is impressive, and we can learn a great deal from it. However, it is not what has traditionally been called "language," and no chimp or gorilla can speak one or, for that matter, more than one language.

In a similar manner, what Boas was doing in 1911 was trying to define human nature at a time when eugenicists were saying that different humans behaved differently because of their genes—that different peoples, races, and even classes behaved differently because of their genetic inheritance. They were just born different. Boas was trying to define what it was to be human, as opposed to others claiming that different people do very different things and that they are born to be different in certain specific ways. Boas was interested in human nature. So, in fact, he was trying to define being human.

He and other anthropologists and social scientists agreed that human beings essentially have certain mental properties that enable them to think in the same way. They have certain specifically human mental processes. However, these mental properties are used, through socialization of the young, to make us grow up with a certain way of looking at the world. Different peoples socialize their young, given their unique tool kit, feeding practices, religion, kinship, social organization, values, and so on, in particular ways, and each child shares certain aspects of his thought and behavior with others in that group. Thus the children of different peoples have different worldviews. They see the world, to some extent, similarly to one another and differently from children brought up in other groups. However, each child, as described by Geertz (1973), begins with all of the natural equipment to live a thousand kinds of human life but in the end having lived only one. Humans have mental properties that allow us each to learn in similar ways but each to learn different things when brought up among different peoples.

Human Nature as Niche Construction

What anthropologists and other social scientist have been doing is trying to explain the end product of what has been happening in human evolution for at least the past two hundred thousand years. We might call this niche construction. Building with a particular body and physiological structure, and with the growth of the nervous system, especially of the brain, there has been an overlap and interaction between the human physical and cultural evolution. Humans are the end product of this organic, psychological, social, and cultural interaction that occurred over this period. As stated by Visala and Fuentes in the introduction to this volume, the boundaries between our genes, epigenetic systems, bodies, ecologies, psychologies, societies, and histories are fluid and dynamic. Perception, meaning, and experience are as central in our evolutionary processes as are nutrients, hormones, and bone density—and all these elements can interact over time.

Essentially, humans created themselves. Humans have culture: they all have language; they can speak; they have and can exchange symbols with one another. However, even though people have language, each culture has only one language. We all understand these things. It is not hard to believe that all people have similar mental processes—unless you are still caught up in the eugenics movement. This is what Boas and other anthropologists and social scientists have been telling us. We realize that other animals can pass on certain learned social behaviors and that they can pass this behavior on to future generations. Animals are smart.

Furthermore, there is no doubt that chimpanzees are extremely adept at learning social skills. A great deal can be learned about how and why this happens by studying other animals. But is what crows, whales and dolphins, capuchins, and chimpanzees do culture? These things (which have been called traditions in ethology) could be called culture, but then a new term is needed for what humans do. We prefer to call what humans do culture and what animals do something else—possibly traditions. In the introduction, Visala and Fuentes said that by exploring the interfaces of cultural and biological evolution, human nature and personality, and the flexibility of the human mind, such a definition of human nature can be found. However, we think that this definition has been with us since 1911.

Culture as Human Nature

We have been attempting to define human culture. It is not easy to do because culture involves so many interactive phenomena. Perhaps it would be better to ask what are some of the things that humans can do and that no other animal is able to do. What do nonhuman animals and humans share with regard to social learning? The ability to learn from one another by observation necessitates one on one, proximate passing on of certain tactile and manipulative tasks. This would be limited to direct observation and proximity in time and space. This would enable animals to pass on some limited learned tasks to one another. In social animals this results in some shared patterns of behavior (relatively limited and stereotyped, usually relatively simple manipulative tasks, like carrying sponges, potato washing, specific hand clasp, choice of different food items). This ability is basically dependent on the proximity of the animals in time and space. Thus, in social animals this can result in relatively simple, though different, patterns of behavior within and between different social groups.

What, though, is unique about humans but also dependent on social learning? One important thing is the ability to think abstract thoughts and the related ability to create and share ideologies and cosmologies. In fact, if we are trying to think of some of the unique tasks that culture enables humans to do we could make a list of these phenomena (and this list, we are sure, is not complete):

Categorizing the world and placing values on categories.

Seeing, understanding, and evaluating the "world."

Passing on ways of "seeing the world"— worldviews, religion, cosmologies, science, and so on.

Asking "why" and "how" questions.

The ability to create alternative worlds; imaginary worlds.

The ability to conceive of ideas about the past and future and things not present in time and space.

The ability to create kinship systems not based on, or limited to, time and space; not dependent on the proximity of particular animals.

The ability to create computational mechanisms (logic) for recursion (structured patterns of thought: mathematics, language, music, and ability to create complex and compound tools and machinery).

The greatly expanded ability to deceive oneself.

The ability to communicate all of the above using abstract symbols not dependent on proximity.

We maintain that these abilities can be defined as culture (and language) and that they are unique to humans. They are empirically observed in all humans, and there is no empirical evidence that they occur in any other animal. In humans these unique behaviors can and do result in profoundly different patterns of behavior within and between different social groups, that is, different worldviews, languages, kinship systems, cultures. However, even though each culture displays different patterns of these unique behaviors, all (normal) humans have the ability to learn and appreciate other cultures and languages. No other animal has this ability. Why? This is because of the biologically based structure of the human mind and the interaction of this mind with our evolution (niche construction) over the past two hundred thousand years. We are not saying that other animals are better or worse; they are different. Just as chimpanzees are different from gorillas, humans are different from chimpanzees and gorillas. Each has a different nature, a different biology, a different biologically based mental structure. Each is unique. We can call what nonhuman animals do "culture" and "language," but then we need other definitions and terms for these things that are unique to humans because otherwise it would trivialize the differences and the uniqueness, and this trivializes the human social sciences (which study this uniqueness and its evolution).

When our youngest daughter was about six or seven, she noticed that there was no number 13 on the buttons of an elevator. She asked what was going on. What happened to floor 13? We told her that it was part of the way we look at the world; some people think that 13 is an unlucky number. She said that's silly because there really was a thirteenth floor. The elevator, the building, the numbers, the lack of 13 on the buttons, the fact that there was a thirteenth floor, and our ability to talk about these things—all are part of what we call culture. No other animal could ask this question. As Clifford Geertz (1973, 49) has said, "Without man, no culture, certainly; but equally, and more significantly, without culture, no men." We would add: Without culture, *no anthropology.* Again, as explained by Geertz:

> We are, in sum, incomplete or unfinished animals that complete or finish ourselves through culture—and not through culture in general, but through highly particular forms of it. . . . We live, as one writer put it, in an information gap. Between what our body tells us and what we have to know in order to function, there is a vacuum we must fill ourselves, and we fill it with information (or misinformation) provided by our culture. (49–50)

Speaking of the process of the evolution of culture over the past two hundred thousand or so years reminds us of one of Gary Larson's cartoons. In it, a number of primitive hominids are interacting with a modern clothed anthropologist (we presume). The caption reads: "Professor Feldman, traveling back in time, gradually succumbs to the early stages of nonculture shock."

References

Balter, M. 2010. "Probing Culture's Secrets: From Capuchins to Children." *Science* 329: 266–67.

Boas, F. 1911a. *Changes in Bodily Form of Descendants of Immigrants.* Reports of the Immigration Commission, vol. 38. Washington, DC: Government Printing Office.

————. 1911b. *The Mind of Primitive Man.* New York: Macmillan.

Boyd, R., and P. J. Richerson. 1996. "Why Culture Is Common, But Cultural Evolution Is Rare." *Proceedings of the British Academy* 88: 77–93.

Degler, C. N. 1991. *In Search of Human Nature: The Decline and Revival of Darwinism in American Social Thought.* New York: Oxford University Press.

de Waal, F. B. M. 1999. "Cultural Primatology Comes of Age." *Nature* 399: 635–36.

Geertz, C. 1973. *The Interpretation of Cultures: Selected Essays by Clifford Geertz.* New York: Basic Books.

Grant, M. 1916. *The Passing of the Great Race.* New York: Charles Scribner's Sons.

Hauser, M. D., N. Chomsky, and W. T. Fitch. 2002. "The Faculty of Language: What Is It, Who Has It, and How Did It Evolve?" *Science* 298: 1569–79.

Helman, C. G. 1994. *Culture, Health and Illness.* Oxford: Butterworth-Heinemann.

Hockett, C. F. 1960. "The Origin of Speech." *Scientific American* 203: 89–97.

Kawamura, S. 1959. "The Process of Sub-Human Culture Propagation among Japanese Macaques." *Primates* 2: 43–60.

Lycett, S. J., M. Collard, and W. C. McGrew. 2007. "Phylogenetic Analyses of Behavior Support Existence of Culture among Wild Chimpanzees." *Proceedings of the National Academy of Science* USA 104: 17588–92.

Marzluff, J. M., and T. Angell. 2005. "Cultural Coevolution: How the Human Bond with Crows and Ravens Extends Theory and Raises New Questions." *Journal of Ecological Anthropology* 9: 69–75.

Nakamichi, M., E. Kato, Y. Kojima, and N. Itoigawa. 1998. "Carrying and Washing of Grass Roots by Free-Ranging Japanese Macaques at Katsuyama." *Folia Primatologica* 69: 35–40.

Perry S., M. Baker, L. Fedigan, J. Gros-Louis, K. Jack, K. C. MacKinnon, J. H. Manson, M. Panger, K. Pyle, and L. Rose. 2003. "Social Conventions in White-Face Capuchin Monkeys: Evidence for Behavioral Traditions in a Neotropical Primate." *Current Anthropology* 44: 241–68.

Rendell, L., and H. Whitehead. 2001. "Culture in Whales and Dolphins." *Behavioral and Brain Sciences* 21: 309–82.

Santorelli, C. J., C. M. Schaffner, and F. Aureli. 2011. "Universal Behaviors as Candidate Traditions in Wild Spider Monkeys." *PloS One.* www.plosone .org/article/info%3Adoi%2F10.1371%2Fjournal.pone.0024400.

Sapolsky, R. 2011. "Warrior Baboons Give Peace a Chance: Born Violent? A Troop of Baboons Chooses an Enduring Culture of Peace." *Yes! Magazine.* www.yesmagazine.org/issues/can-animals-save-us/warrior-baboons-give -peace-a-chance.

Spiro, J. P. 2009. *Defending the Master Race: Conservation, Eugenics, and the Legacy of Madison Grant.* Burlington: University of Vermont Press.

Sussman, R. W. 2014. *The Myth of Race: The Troubling Persistence of an Unscientific Idea.* Cambridge, MA: Harvard University Press.

Thornton, A., and K. McAuliffe. 2006. "Teaching in Wild Meerkats." *Science* 313: 227–29.

Van Schaik, C. P., M. Ancrenaz, G. Borgen, B. Galdikas, C. D. Knott, I. Singleton, A. Suzuki, S. S. Utami, M. Merrill. 2003. "Orangutan Cultures and the Evolution of Material Culture." *Science* 299: 102.

Whiten, A., J. Goodall, W. C. McGrew, T. Nishida, V. Reynolds, Y. Sugiyama, C. E. G. Tutin, R. W. Wrangham, and C. Boesch. 1999. "Cultures in Chimpanzees." *Nature* 399: 682–85.

CHAPTER 2

"TO HUMAN" IS A VERB

TIM INGOLD

I

The time is July 1885, the place Mount McGregor, to which the eighteenth president of the United States, Ulysses S. Grant, has retired to write his memoirs. On his deathbed, unable to speak because of the throat cancer that was killing him, Grant penciled the following note to his doctor, John H. Douglas: "The fact is I think I am a verb instead of a personal pronoun. A verb is anything that signifies to be; to do; or to suffer. I signify all three."[1] There is no knowing what exactly was going through Grant's mind as he wrote these gnomic lines, for he died a few days later. My purpose in this essay, however, is to offer some reflections on what he might have meant, for I believe that his words encapsulate a profound solution to what is surely the oldest and most fundamental problem of anthropology: what, exactly, does it mean to think of ourselves that we are human?

More than five hundred years earlier, on the island of Majorca, the same problem was exercising the mind of the writer, philosopher, and mystic Ramon Llull.[2] Born in 1232 to an aristocratic family, and by his own account, Llull lived the dissolute life of the troubadour until one day, while composing a love song to his latest paramour, a vision came to him of Christ suspended on the cross. Over subsequent days the vision kept recurring, causing him such alarm that he eventually resolved to abandon his licentious ways and devote the rest of his life to Christian teaching and scholarship. At that time, Majorca was a center of commerce in the Mediterranean world and a melting pot of ideas from Islam, Judaism, and Christianity. Realizing that to convince Moslems and Jews of the truth of Christianity meant approaching the subject in an ecumenical spirit, Llull embarked on nine years of intense study, including learning Arabic from a Moslem slave he had purchased but with whom he subsequently fell out (imprisoned for blasphemy, the Saracen eventually hanged himself in jail, saving Llull from the awful responsibility of having to decide on his fate). This study laid the foundations for an extraordinarily long and prolific life, during which he wrote some 280 books, composed in Latin and Arabic as well as in his native Catalan. One of the last of these was the *Logica Nova*, written in Genoa in 1303, in his seventy-first year.

Much inspired by his engagements with Islamic culture and science, Llull presents us in this work with a dynamic cosmos in which everything there is—every entity or substance—is what it is thanks to the activity proper to it. Things, for Llull, are what they do. For example, it is of the essence of fire that it burns. Precisely what fuels the fire, or what is heated by means of it, is an accidental or contingent matter. Perhaps you burn wood to heat water, but neither wood nor water is necessary for there to be fire. What *is* necessary is that burning should be going on. Likewise, whiteness may whiten this or that body, but there is only whiteness when whitening is going on (Lohr 1992, 29–30). That the existence of a thing or substance is indistinguishable from its activity is not, however, easily expressed in Latin, and in order to achieve this Llull had to devise new words, modeled on the forms of the Arabic verbs with which he was familiar. One of these neologisms appears when he turns to the problem of defining the human. If what goes for everything else goes for human beings too, then they must likewise be defined by the activity proper to them. Where there are humans, something must be going on. But what? Once again, Llull had to invent a new verb: *homificare*, 'to humanify'. The human, according to Llull's enigmatic definition, is a humanifying ani-

mal: *Homo est animal homificans.*[3] Precisely what human beings do, or
how they go about it, is by the way. However, wherever and whenever
they exist, humanifying is going on. Humans humanify themselves, one
another, the animal and vegetable kingdoms, and indeed the entire universe (Lohr 1992, 34).[4] Thus for Ramon Llull, nearing the end of his long
life, as indeed for Ulysses Grant over five centuries later, it seemed that
the grammatical form of the human is not that of the subject, whether
nominal or pronominal, but that of the verb.

Hard as this may have been to express in a language that normally
enlists the verb into the predicate, and thus categorically separates persons
and things, as causal agents, from the effects they set in train, it would
have come easily to those peoples who are credited, in current anthropological parlance, with ontologies of animism—that is, ontologies in which
to exist *means* to participate from within in the continual unfolding of a
lifeworld. Precisely such an understanding of existence is current among
the people of the Kelabit Highlands of central Borneo, as described by the
anthropologist Monica Janowski (2012). For the Kelabit, the cosmos and
everything included in it is infused with a vital energy known as *lalud*. As
it flows and percolates through the entire landscape, so *lalud* brings everything to life. The cosmos, as Janowski puts it, "is dynamically animated"
(148). But if *lalud* is everywhere, it nevertheless varies in its concentra-
tion. As it concentrates, it coalesces into entities and substances of mani-
fold kinds, including humans and animals, rocks and trees. The older
things are, the more *lalud* is concentrated in them and the harder or more
solid they become. At the same time, however, they become more powerful, and this power is manifested in their capacity to interrupt or chan-
nel the current of *lalud* itself. In this regard, human beings are particularly
powerful. Kelabit distinguish between the bare fact of living, *mulun*, and
the way of life, *ulun*, peculiar to the human (156). In the practice of *ulun*,
humans do not merely go with the flow, as other creatures do. Rather,
they intervene in directing it, bending or diverting it in ways that answer
to their purposes, above all to the imperative to provision themselves.
Principally this is done through the construction of earthworks, including
irrigation and drainage ditches, and even cutting through river meanders
to encourage the formation of oxbow lakes, all with the aim of creating
wet fields for rice cultivation. Really powerful people are distinguished

from lesser folk, and human beings more generally from animals, by the enduring marks, or *etuu*, that their *lalud*-directing accomplishments have left in the landscape.

II

As world shapers whose very existence and capacities are congealed within currents of life activity that they themselves aspire to direct, human beings are, for the Kelabit, precisely what Llull also took them to be, namely, humanifying animals. They do not set out, as an ontology more conventional to the Western tradition would have it, to *humanize* the world, that is, to superimpose a preconceived order of their own on a given substrate of nature. For humans to humanify, in the sense that Llull intended and that corresponds with the Kelabit idea of *ulun*, is to forge their own existence within the crucible of a common lifeworld. Their humanness is not given from the start, as an a priori condition, but emerges as a productive achievement—one, moreover, that they have continually to work at for as long as life goes on, without ever reaching a final conclusion. Or in the words of the twentieth-century Spanish philosopher José Ortega y Gasset (1961, 200): "The only thing that is given to us and that *is* when there is human life is the having to make it, each one for himself. . . . Life is a task" (original emphasis). This passage comes from a celebrated essay titled "History as a System," which Ortega composed in 1935, just prior to the outbreak of the Spanish Civil War, while living in exile in Buenos Aires, Argentina. Echoing Llull, Ortega argues that the grammatical form of life is that of the gerund: it is always in the making, "a *faciendum* not a *factum*" (200). For that reason, he thinks, appeals to human nature or, alternatively, to the human spirit are misconceived. To speak of the human body or the soul, or of the psyche or spirit, is to suppose that such a thing has already crystallized out, in a fixed and final form, from the processes that gave rise to it. It is to place, at the origin, a conclusion that is never actually reached. For in truth, where there is human life there is never anything but happening. Life *is* not; it *goes on*. Indeed, as Ortega observes, there is a certain absurdity in our customary way of referring to ourselves, as human beings (213). For how can one go on being? It is like asking us to move along and stand in one place at the same time.

Perhaps, then, we should substitute the word *becoming* for "being."
As instantiations of life-in-the-making, should we not rather call ourselves
human becomings? In an intriguing aside, Ortega (1961, 200 n. 9) rules
out such an alternative, with critical reference to an earlier philosophical
writer with whom he is otherwise very much in sympathy. That writer
was Henri Bergson. For Bergson, too, it was all happening. Everything
was movement, growth, becoming: the apparently fixed forms of things
but the envelopes of vital processes. Being, said Bergson, lies in self-
making: *l'être en se faisant*. Yet in Bergson's vocabulary, self-making was
just another word for becoming (from *devenir*, 'to become'). Ortega insists, to the contrary, that there is more to the human task of self-making
than mere becoming; more to life-making than mere living; more, as the
Kelabit would say, to *ulun* than *mulun*. Humans are quite literally the
fabricators of themselves; they are *auto-fabricators* (Ortega y Gasset 1961,
115). Indeed in an argument strikingly redolent of Kelabit reflections on
the same theme, Ortega claims that unlike other animals, which merely
become whatever it is in their nature to be, humans must perforce determine what they are *going* to be. The fulfillment of human being is always
deferred, always not yet: "man," says Ortega, is a "not-yet being" or, in a
word, an "aspiration" (112–13). And precisely because they aspire to
things, humans also face difficulties in their achievement (201). Life is
not difficult for the animal, since it does not reach out for what is not immediately attainable. Nor, for that matter, is it easy. The difference between ease and difficulty is of no concern to the animal. But for humans,
caught as they are between the reach of aspiration and the grasp of prehension, it is a never-ending preoccupation.

To put it another way, by comparison to the animal in whose horizon
there is no past or future, only an ever-evolving now, the movement of
human life is temporally stretched. Out in front is the "not yet" of aspiration, bringing up the rear the "already there" of prehension. At once
not yet and already, humans—we might say—are constitutionally ahead
of themselves. Whereas other creatures must be what they are in order to
do what they do, for humans it is the other way around. They must do
what they do to be what they are. You must be a bird to fly, but to be
human, you must speak. Flying does not make a bird, but speaking makes
us human. It is not that humans are becoming rather than being; rather
their becoming is continually overtaking their being. This, I suggest, is

also what the Kelabit have in mind when they distinguish the *ulun* of human endeavor from the *mulun* or bare life of other creatures. And it is what Llull had in mind when he spoke of man as a humanifying animal. Moreover, I think it is probably at the back of the minds of most of us when we say of our human selves that we do not just live our lives, but *lead* them. What Llull's humanifying, Kelabit *ulun*, and Ortegan auto-fabrication have in common, then, is that they are all about leading life. As an answer, however, this merely kicks the question down the road. The question was, what does it mean to think of ourselves that we are human? All we have managed to do so far is to replace this with another question, namely, what does it mean to say of lives that they are led? The answer I propose, in what follows, is that to lead life is to *undergo an education*.

III

Both terms need to be unpacked, and I shall start with the verb *to undergo* before turning to the significance of *education*. In a work titled *Intellectual Foundations of Faith*, dating from 1961, the American theologian Henry Nelson Wieman set out to show in what ways a life that is led—that is to say, a *human* life—can be creative. It is necessary, he argued, to distinguish between two kinds, or meanings, of creativity (Wieman 1961, 63–66). There is, on the one hand, the creativity that is expressed in what people *do*. A person is creative in this sense when he or she "constructs something according to a new design that has already come within reach of his [or her] imagination." This is the sense most commonly invoked when creativity is identified with innovation. It is found by looking back from a final product—what Wieman calls a "created good"—to an unprecedented idea in the mind of an agent, in whose doing or making it was actualized. On the other hand, however, is the creativity that "progressively creates personality in community."[5] Wieman's point is to argue that behind the contingencies of what people do, and the miscellany of products or created goods to which these doings give rise, is the "creative good" that is intrinsic to human life itself, in its capacity to generate persons in relationships. This kind of creativity, he says, is "what personality undergoes but cannot do" (Wieman 1961, 65–66). It does not begin here, with an idea in mind, and end there, with a completed object. Rather it carries on through, without beginning or end. Such is the creativity of social life. For social life is not something that the person does but what the person undergoes: a process in which human beings do not create societies but, living socially, create themselves and one another. That is to say, they both grow and are grown, undergoing histories of development and maturation—from birth through infancy and childhood into adulthood and old age—within fields of relationships established through the presence and activities of others. And critically, this growth is not just in strength and stature but also in knowledge, in the work of the imagination and the formation of ideas.

As a young man, Wieman had been a keen reader of the philosophical writings of Henri Bergson, and he would later go on to study and publish on the work of Bergson's British contemporary, Alfred North Whitehead. It was Whitehead (1929, 410) who coined the term *concrescence* to describe the capacity of living things continually to surpass themselves. In a world of life, Whitehead argued, there are not only concrete, created things but also concrescent, crescent things. Or rather, one could look at this same world in two ways, either from the outside, considering every organism as the living embodiment of an evolved design, or from the inside, by joining with the generative movement of its growth and formation—that is, of its coming into being, or *ontogenesis*. Wieman's distinction between the creativities, respectively, of doing and undergoing, or between created goods and the creative good, is clearly a version of the same thing. Moreover, there are echoes of Bergson in the idea of a creativity that is to be found not in the characteristic doings of the person but in the creation of personality in community. "It is . . . right to say that what we do depends on who we are," wrote Bergson in his *Creative Evolution* of 1911, "but it is necessary to add also that we are, to a certain extent, what we do, and that we are creating ourselves endlessly" (7). This endless creation of ourselves corresponds precisely to Wieman's idea of a creativity undergone rather than done. Furthermore, as Bergson was keen to stress, the process is irreversible. Thus to understand creativity in this sense is to read it forward, in the unfolding of the relations and processes that actually give rise to worldly beings, rather than back, in the retrospective attribution of final products to initial designs. It is to recognize, with

Bergson, that ontogenesis takes time. This is time as duration: not a succession of instants but the prolonging of the past into the actual. "Duration," Bergson (1911, 4–5) wrote, "is the continuous progress of the past which gnaws into the future and which swells as it advances."

But is this idea of gnawing and swelling sufficient to grasp the essence of lives that are not just lived, but led? The past might gnaw into the future, but the future has a habit of slipping away beyond our reach. We have to track it down. For Ortega y Gasset, as you will recall, what is distinctive about human life is that it is not just impelled from behind but summoned from in front. Humans, he argued, are the scriptwriters or novelists of their own lives, creating themselves as they go along. As every novelist knows, characters have a way of outrunning their author's capacity to write them down. It is vital not to lose them. So, too, in the creation of our own lives, we are fated to give chase to hopes and dreams that are forever on the point of vanishing (Ingold 2013, 71–72). And since all human life is happening, so all creation is occasional: a moment-to-moment improvisation. Whereas God created the world in a single act, and finished the job, "man," wrote Ortega, "makes himself in the light of circumstance, . . . he is a God as occasion offers, a 'secondhand God'" (Ortega y Gasset 1961, 206). And it is precisely in this task of secondhand creation that imagination comes into play. God needs no imagination, since his creation is already all in place before the act begins. But worldly, mortal humans can only re-create piecemeal, a bit at a time. Recall that in Wieman's definition, "to do" is to act on an idea that is already settled within the imagination. The intention or design is there before the act. "To undergo," by contrast, is to move upstream, to where ideas have yet to crystallize out from the flow of action. As a creative good, the power of the imagination is not one of mental representation, nor is it a capacity to construct images in advance of their material enactment. It is rather the generative impulse of a life that continually runs ahead of itself. Following Ortega, we could say that *imagination* is another word for the aspiration of not-yet-being. As such, it leads from the front rather than pressing from behind. But *where* it leads is not yet plotted out before the act begins. As we say colloquially, the propensity of the imagination is to roam, to cast about for a way ahead or to improvise a passage; it is not to map final outcomes and all the steps to reach them. And for Ortega,

without imagination—without this capacity to run ahead of ourselves—
human life would be impossible.

IV

This leads me to my second term, namely, *education*. Like the concept
of creativity, the word *education* carries a double connotation. The first
is familiar enough to all of us who have sat in a school classroom as pu-
pils or who have stood up before the class to teach. This is the sense of
the Latin verb *educare*, meaning "to rear or bring up," to instill a pattern
of approved conduct and the knowledge that supports it. The efficacy
of education, here, lies in the intergenerational transmission of an order
that already exists. It is, according to the normative canon of Western
pedagogy, the provision of cultural content to young minds equipped by
nature with the capacities to receive it. As such, it is a process of humani-
zation. A variant etymology, however, traces the word to *educere*, from *ex*
(out) plus *ducere* (to lead). In this sense, education is a matter of leading novices *out* into the world rather than—as it is more conventionally
understood—of instilling knowledge *in* their minds. To be thus led out
is to be both impelled and drawn into the unknown. Education, in this
second sense, is not what a person receives in order that they may then
go on to do this or that in accordance with existing social or moral pre-
cepts. It is rather what the person undergoes: a process not of humaniza-
tion but of humanifying. It might perhaps be compared to going out for
a walk. In a brilliant meditation on the subject of walking and education,
the philosopher Jan Masschelein (2010a, 2010b) argues that walking is
not about taking up a standpoint on things. It does not place us in a position, or afford us a perspective, on the basis of which we can then proceed to act. On the contrary, walking continually pulls us away from *any*
standpoint—from any position or perspective we might adopt. "Walking," as Masschelein (2010b, 276) explains, "is about putting this position at stake; it is about ex-position, about being out-of-position." Or in
a word, it is a practice of *exposure*. And this is precisely what is achieved
by education in the sense of *ex*-duction.

Crucially, the movement of ex-duction, of a life that leads out into a
world-in-formation, is not intentional but *attentional*. When we say of a

deed that it is done with intention, we mean that the outward cast of action follows an inward cast of thought. The mind intends and the body extends. This is what Wieman implied in his definition of "doing" as action predicated upon a design already lodged within the mind of the agent. Yet to the extent that the doer is wrapped up in the seclusion of his own deliberations, he is perforce absent from the world itself. On exposure, however, this protective wrapping unravels. The walker on the move, lest he lose the way, must be ever vigilant to the path as it unfolds before him. He must watch his step, and listen and feel as well. He must, in a word, pay attention to things, and adjust his gait accordingly. Thus in the very act of walking, he is thrust into the presence of the real. As intention is to attention, therefore, so absence is to presence. A person might intend to go for a walk; she might reflect on it, consider the route, prepare for the weather, and pack her provisions. In that sense walking is something she sets out to do. She is the subject, her walking the predicate. But once on the trail, she and her walking become one and the same. And while there is of course a mind at work in the attentionality of undergoing, just as there is in the intentionality of doing, this is a mind immanent in the movement itself rather than an originating source to which such movement may be attributed as an effect. Or in short, if the walker's intention converges on an origin, her attention comes from being pulled away from it—from displacement.

At first glance this conclusion seems remarkably close to that reached by the ecological psychologist James Gibson (1979). Pioneering his ecological approach to visual perception, Gibson had proposed that we do not perceive our surroundings from a series of fixed points; nor, he argued, is it the task of the mind to assemble, in memory, the partial perspectives obtained from each point into a comprehensive picture of the whole. Rather, perception proceeds along a *path of observation*. As the observer goes on his way, the pattern in the light reaching the eyes from reflecting surfaces in the environment (i.e., the "optic array") is subject to continual modulation, and from the underlying invariants of this modulation, things disclose themselves for what they are. Or more precisely, they disclose what they afford, insofar as they help or hinder the observer to keep on going or to carry on along a certain line of activity. The more practiced we become in walking these paths of observation, according to

Gibson, the better able we are to notice and to respond fluently to salient aspects of our environment. That is to say, we undergo what Gibson (1979, 254) called an "education of attention." Many scholars, myself included (Ingold 2000, 166–68), have followed this approach in describing the process of enskillment by which the novice is gradually transformed into a "master" of what he or she does. Walkers become skilled in detecting and responding to irregularities of the ground surface, enabling them to keep their balance in difficult terrain. Musicians become skilled in the manipulation of their instruments, enabling them to find their way through the most complex passages. Weavers become skilled in working with fibrous materials, enabling them to generate a regular pattern from the rhythmic repetition of basic movements.

Yet with mastery comes its opposite: submission. To embark on any venture—whether it be to set out for a walk, to play a piece of music, or to weave a textile—is to push one's boat out into the stream of a world in becoming, with no knowing what will transpire. It is a risky business. Thus when Masschelein describes walking as a practice of exposure, both the education to which the walker lays himself open and the attention demanded of him are quite the reverse of what Gibson had in mind in his theory of perceptual attunement. It is not a matter of picking up, and turning to one's advantage, the affordances of a world that is already laid out. Recall that the verb *attendre*, in French, means "to wait," and that even in English, to attend to things or persons carries connotations of looking after them, doing their bidding, and following what they do. In this regard, attention abides with a world that is not ready-made but always incipient, on the cusp of continual emergence. In short, whereas in one sense, in the acquisition and exercise of skilled mastery, the world waits for the practitioner, in another sense, in the educational practice of exposure, the practitioner waits upon the world. Here, to wait—to attend—is to be commanded not by the given but by what is *on the way* to being given (Masschelein 2010a, 46). Thus the walker, a master of the terrain, must wait for signs that reveal the path ahead, with no surety of where it will lead; the hunter, a master of the chase, must wait for the animal to appear, only to put himself at risk in its pursuit; the mariner, a master of his ship, must wait for a fair wind, only to submit to the elements. The walker, as indeed the hunter and the mariner, once embarked upon a course, is at the mercy of the befalling of things. In these as in countless other examples, mastery and vulnerability, practical enskillment and existential risk, are two sides of the same coin. That coin is attention.

V

What, then, is the relation between the two sides: between our waiting for the world and the world's waiting for us and between the modes of education that lie, respectively, in exposure and in attunement? Earlier, I suggested that unlike other creatures that live their lives but do not lead them, the lives of humans are temporally stretched, between the already and the not yet. It seems that in every venture and at every moment, we are both fully prepared and yet utterly unprepared for things to come. What then leads, and what follows? The usual answer is to claim that as intentional beings—that is, as agents—humans deliberate before they act, in Wieman's sense of doing what has already come within reach of the imagination. Thus the mind commands and the body submits more or less mechanically to its directions. Mastery, in this account, is cognitive: if humans lead their lives it is entirely thanks to their capacity to conceive of designs in advance of their execution, something of which animals— at least for a science of mind constructed on Cartesian principles—are deemed incapable. The chess master, for example, plans his moves in his head, by means of mental computations of wondrous complexity, whereas their subsequent enactment, entailing the grasping and lifting of a piece from one square and its transport to another, could hardly be simpler. It requires no great skill; indeed any machine could do it. I would like to propose, however, that in undergoing, the relation of temporal priority between mastery and submission is the reverse of that which is assumed in the cognitive or intentionalist account of doing. Here, *submission leads and mastery follows:* education as exposure precedes education as attunement. Rather than a commanding mind that already knows its will trailing a subservient body in its wake, out in front is an aspirant imagination that feels its way forward, improvising a passage through an as yet unformed world, while bringing up the rear is a prehensive perception already accustomed to the ways of the world and skilled in observing and responding to its affordances.

A life that is led, then, or one that undergoes an education, is held in the tension between submission and mastery, between imagination and perception, between aspiration and prehension, and between exposure and attunement. In every one of these pairings, the first leads and the second follows. But the former's lead is not commanding but tentative. It requires of its following not passive obedience but active delivery. Pushing the boat out, I call upon my powers of perception to respond. Yet in that very response I discover that unbeknownst to me, I have been there before, as have my predecessors since time immemorial. Without even thinking about it, I seem to know the ropes. Heading out along the trail, into the "not yet," I already know how it goes. Thus *all undergoing is remembering*. As the phenomenologist Bernhard Waldenfels (2004, 242) has put it, "we are older than ourselves": behind the selves we are on the point of becoming, but are not yet, are the selves that we already are without our knowing. In this ongoing, iterative process of becoming who we were, and of having been whom we become, there is no bottom line, no point at which we can uncover some basic human nature that was there before it all began. In their animic cosmology, the Kelabit remind us that humanness is not given a priori but is a productive achievement that leaves its marks in the landscapes of habitation (Janowski 2012, 154–56). Or as the anthropologist Istvan Praet (2013, 203) has recently written, with regard not specifically to the Kelabit but to ontologies of animism from around the world, "What always recurs . . . is a Humanity envisaged as an ongoing fabrication rather than a guaranteed status, . . . a sustained pattern of activity or particular way of living rather than . . . an absolute class or category." Remember what Ortega said: we are secondhand gods, not created once and for all, but creating and re-creating ourselves as the occasion demands. Thus as an *animal homificans*, in Llull's phrase, I am my walking, and my walking walks me. So here's a riddle: I carry on and am in turn carried. I live and am lived. I am both younger and older than myself. What am I? President Ulysses Grant was right. I think I am a verb.

I would like to conclude with a poem. It is by the philosopher Jean-Luc Nancy, and its title is *The Instructions*. The poem was prominently displayed on a large glass panel as part of the *Do It* exhibition held at the Manchester Art Gallery from July to September 2013. The exhibition provided its public with dozens of instructions, ranging from the active to the absurd, which visitors could try out for themselves, either in the gallery or at home. Judged as a poem, *The Instructions* is perhaps proof that poetry is better left to poets than to philosophers. However, it happens to encapsulate almost everything I have argued that doing is not. Indeed, in the poem Nancy is inviting us to think of doing in a way to which we are quite unaccustomed. And that way corresponds almost exactly to what I, following Wieman, have called undergoing.

> Do it!
> 'it': What you have to do,
> What is up to you to do,
> What falls to you
>
> 'it': Undetermined, undeterminable,
> Which will only exist when you have done it
>
> Do it, do that,
> That thing no-one expects,
> Not even you,
> That improbable thing
>
> Do what stems from your doing,
> And yet is not done by you
> Nor produced
> But stems from well before your doing
> From well before you
>
> Do what escapes you
> What is not yours
> And that you owe.

First of all, says Nancy, the "it" that you "do" is not already within reach, ideally if not materially, before you start. Doing, in other words, does not translate from an image in the mind to an object in the world. Rather, both the thing and the idea of it emerge together from the doing itself. This doing, moreover, is an act to which you submit: you do not initiate it; rather it *falls* to you. It was the last thing you expected to happen, and in undertaking the task, you were perhaps surprised to discover capacities of perception and action you never knew you had. But where has it come from, this thing you did? For Nancy, it has no point origin; it cannot be traced to an intention. What we do is not done by an authorial agent with a design in mind. It is rather part of a never-ending process of attention and response in which, as we have seen, all human life is caught. Just as the "already" is always behind us, as far back as we care to go, so the "not yet" will always escape ahead of us, beyond the horizon of our expectations. And as we owe our very existence to what has gone before, and as what comes after owes its existence, at least in part, to us, so our deeds belong to no one: not to ourselves, not to others, but to history—or better, to life. The doing of which Nancy speaks—the doing that is not done by you—is a kind of action without agency, an undergoing, an auto-fabrication. It is not *mulun* but *ulun*, as the Kelabit would say: a life not just lived, but led. It is a humanifying. But let us drop such obfuscating words and simply end where we began: for us, "to human" is a verb.

Acknowledgments: This essay owes its conception to a memorable conversation that took place during the 17th World Congress of the International Union of Anthropological and Ethnological Sciences (IUAES) at the University of Manchester, August 5–10, 2013, with Thomas Schwarz Wentzer and the members of his Philosophical Anthropology Group (Line Ryberg, Kasper Lysemose, Ramus Dyring) from the University of Aarhus, Denmark. I particularly want to thank Thomas Wentzer for drawing my attention to the works of Ramon Llull. At the subsequent International Human Research Conference, which took place at Aalborg University (August 13–16), and at which I had spoken on the topic of human creativity, a questioner asked me to clarify the relation between doing, undergoing, and agency. This essay is my provisional answer.

Notes

1. The semiotician Thomas A. Sebeok (1986, 1–2) gives an account of this episode in the introduction to his collection of essays, *I Think I Am a Verb*.

2. For details of Llull's life and work, I have drawn on the authoritative works of Anthony Bonner (1985) and Charles Lohr (1992).

3. Here I follow Bonner's translation: "man is a manifying animal" (in Llull 1985, 609).

4. Lohr renders the verb *homificare* as "hominize" rather than "humanify." I explain my preference for the latter below.

5. For further discussion of this distinction, see Ingold 1986, 202 ff.; and Ingold and Hallam 2007, 8.

References

Bergson, H. 1911. *Creative Evolution*. Trans. A. Mitchell. London: Macmillan.

Bonner, A. 1985. "Historical Background and Life of Ramon Llull." In *Selected Works of Ramon Llull (1232–1316)*, vol. 1, ed. and trans. A. Bonner, 5–52. Princeton, NJ: Princeton University Press.

Gibson, J. J. 1979. *The Ecological Approach to Visual Perception*. Boston: Houghton Mifflin.

Ingold, T. 1986. *Evolution and Social Life*. Cambridge: Cambridge University Press.

———. 2000. *The Perception of the Environment: Essays on Livelihood, Dwelling and Skill*. London: Routledge.

———. 2013. *Making: Anthropology, Archaeology, Art and Architecture*. Abingdon: Routledge.

Ingold, T., and E. Hallam 2007. "Creativity and Cultural Improvisation: An Introduction." In *Creativity and Cultural Improvisation*, ed. E. Hallam and T. Ingold, 1–24. Oxford: Berg.

Janowski, M. 2012. "Imagining the Forces of Life and the Cosmos in the Kelabit Highlands, Sarawak." In *Imagining Landscapes: Past, Present and Future*, ed. M. Janowski and T. Ingold, 143–63. Farnham: Ashgate.

Llull, R. 1985. *Selected Works of Ramon Llull (1232–1316)*, vol. 1, ed. and trans. A. Bonner. Princeton, NJ: Princeton University Press.

Lohr, C. 1992. "The New Logic of Ramon Llull." *Enrahonar* 18: 23–35.

Masschelein, J. 2010a. "E-ducating the Gaze: The Idea of a Poor Pedagogy." *Ethics and Education* 5(1): 43–53.

———. 2010b. "The Idea of Critical E-ducational Research—E-ducating the gaze and Inviting to Go Walking." In *The Possibility/Impossibility of a New Critical Language of Education*, ed. I. Gur-Ze'ev, 275–91. Rotterdam: Sense Publishers.

Ortega y Gasset, J. 1961. *History as a System and Other Essays toward a Philosophy of History*. New York: W. W. Norton.

Praet, I. 2013. "Humanity and Life as the Perpetual Maintenance of Specific Efforts: A Reappraisal of Animism." In *Biosocial Becomings: Integrating Social and Biological Anthropology*, ed. T. Ingold and G. Palsson, 191–210. Cambridge: Cambridge University Press.

Sebeok, T. A. 1986. *I Think I Am a Verb: More Contributions to the Doctrine of Signs*. New York: Plenum Press.

Waldenfels, B. 2004. "Bodily Experience between Selfhood and Otherness." *Phenomenology and the Cognitive Sciences* 3: 235–48.

Whitehead, A. N. 1929. *Process and Reality: An Essay in Cosmology*. Cambridge: Cambridge University Press.

Wieman, H. N. 1961. *Intellectual Foundations of Faith*. London: Vision Press.

RESPONSE I

Free and Easy Wandering

*Humans, Humane Education, and Designing in
Harmony with the Nature of the Way*

SUSAN D. BLUM

Almost thirty years ago I wrote a master's thesis in Chinese literature titled "Of Motion and Metaphor: The Theme of Kinesis in *Zhuangzi.*"
Somewhere in the many boxes that a packrat academic has accumulated
is a copy of that thesis. I did find the penultimate version in WordStar on
a 5-inch floppy disk.

But what I want to talk about is not hoarding or the transformation
of communicative media.

I want to talk about motion.
And education.
And human nature(s).

And, in the context of the human nature project, developed over the last four years or so, where it was made clear that to limit an almost-limitless-goal, the focus would stick with Western thought, I am here to add ancient China to the conversation. As Tim Ingold notes, anthropology's contribution is often simply to lend some exotic expertise to the conversation.

But I add China not just as "another country heard from," in Geertz's half-dismissive and half-admiring phrase. I add China because more than two thousand years ago the tension between humanity as a verb (Ingold 2014) and humanity as a substance was already being sorted out, with competing views enduring ever since. And because the beauty of the exposition can be an antidote to the dehumanizing abstraction of much philosophical discussion, I turn to the *Zhuangzi*. I want to weave together the *Zhuangzi*, education, evolution (Bock 2010), human nature(s), language, permaculture, walking, and lines (Ingold 2007)—in my somewhat less than three thousand words.

I

The book *Zhuangzi*, one of the two core texts of philosophical Daoism, was compiled from writings of the fourth, third, and second centuries BCE. It is especially known for the parable of Zhuangzi dreaming of being a butterfly—or is it the butterfly dreaming of being Zhuangzi?

The mystical opening chapter of the book is translated variously as "Free and Easy Wandering" or "Going Rambling without a Destination." It includes a dizzying portrayal of movement, motion, and transformation.

> In the northern darkness there is a fish and his name is *kun*. The *kun* is so huge I don't know how many thousand *li* he measures. He changes and becomes a bird whose name is *peng*. The back of the *peng* measures I don't know how many thousand *li* across and, when he rises up and flies off, his wings are like clouds all over the sky. When the sea begins to move, this bird sets off for the southern darkness, which is the Lake of Heaven.

The *Universal Harmony* records various wonders, and it says, "When the *peng* journeys to the southern darkness, the waters are roiled for three thousand *li*. He beats the whirlwind and rises ninety thousand *li*, setting off on the sixth month gale." Wavering heat, bits of dust, living things blowing each other about—the sky looks very blue. Is that its real color, or is it because it is so far away and has no end? When the bird looks down, all he sees is blue too. (*Zhuangzi* chap. 1, "Free and Easy Wandering," trans. Watson)

On it goes, delighting and confusing and enlightening its willing readers.

Zhuangzi's Daoism has been contrasted for more than two thousand years with the uprightness and squareness, the being *zheng*, of the Ruists (known as Confucians). For Ruists, all was square corners and straight lines. For Zhuangzi, all was curves, sauntering, riding the air currents, spiraling. This is what it meant to be at one with nature, because only by following nature, the Dao, and its principles could one achieve ease. If a person, or an animal, were to do this, there would be no tension. Just as a knife never needs sharpening when wielded by an expert carver who knows how to slip it between the joints of an animal being butchered, so a person wandering in the way can live ten thousand years.

And while the Ruists had their commentaries and their masters, and later on their examinations—which gave the world the idea of civil service examinations (Elman 2000) and Tiger Mothers (Chua 2011)—the Daoists learned how to be harmonious from the cats that loped around and from the water that flowed.

Many have wondered if the Chinese philosophical tradition emphasizing action rather than being or essence stems from the lack of the copula, the verb *to be*, in contrast to ancient Greek, which reveled in the conjugations and cases of *einai* 'to be' (Benveniste [1958] 1971; Kahn 1966; Sivin and Lloyd 2002).

Others explain Chinese thought as stemming from the apparent concreteness of the written form of Chinese (Fenollosa and Pound [1919] 2008; but see Saussy 2008; DeFrancis 1984). In my little thesis, I explored the semantic classifier indicating movement, which looks like this: 辵 or 辶. I took it as a metaphor for the focus on motion pervading this text.

II

The same tension, the tension between preset outcomes and delighted trust in serendipitous results, is being explored in permaculture, the set of design principles that aims to work entirely with natural laws to create mutually reinforcing entities that are quite nearly self-sustaining (Holmgren 2002). The most evident contrast is with the straight lines of industrial agriculture, made possible by million-dollar combines, harvesters, and highway subsidies, with tasteless vegetable-like items that look like natural substances and are shipped thousands of miles.

And the contemporary education system, what I call industrial education (Blum 2013, 2016), similarly begins with a fixed model, a destination, a set of procedures all specified in advance. All that human substance—the humanifying—has to be forced into the mold and shoved out at the end. There may have once been a sense of movement and diverse shapes, but by the end we find lifeless human sausages (Henry 1963; Holt [1976] 2004; Illich 1970).

Except that the human spirit is too strong for that, so students find ways to reject this uniform stamping of their being. We find all kinds of resistance and failure; such uniformity is impossible.

The "carpentered world" hypothesis states that our perception is shaped by expectations from the built environment. It accounts for certain optical illusions (e.g., the Müller-Lyer illusion; Segall, Campbell, and Herskovits 1966) and our use of perspective in painting. But it also shapes expectations for aesthetics, for beauty, for what is right and normal in the world (Henrich, Heine, and Norenzayan 2010). In the industrial classroom attention must not wander. Focus, order, efficiency, a direct path, and motionlessness, as the goal is kept in mind, must prevail.

In alternative forms of education, if they occur in a school, much dis-order and even directionlessness are permitted. Forms of "traditional" education insist that children need to be kept in line, orderly, still. A well-ordered classroom is quiet and motionless. "Progressive" education, building on the images offered by Rousseau ([1762] 1979), sees a trustworthy "nature" in children and aims to follow that nature, to get out of the way and allow the child free and easy wandering, rambling without a destination (Dewey 1966; Edwards et al. 2013; Stevens 2001).

The various movements of homeschooling and unschooling, imaginative education that proliferates, now vary about whether there is a destination, or rather the possibility of surprise. This cannot be tested on standardized tests. And this view of children is as far from the industrial widget version as possible.

III

It is evident that there is no single human nature. We know that variation is not only possible, but inevitable. So if in permaculture the aim is to work with natural processes to encourage harmonious and stress-free growth, how do we do that in education, when there are competing views of the nature of humans?

Are children by nature wise or evil? The Chinese philosophers have debated this too for millennia, and anthropologists dredge up samples of societies that have competing views. David Lancy (2014) contrasts views of children as "chattel, changelings, or cherubs." Whatever the view of humans, and the younger ones we often call children, this determines our educational models.

And should education work against or with natural tendencies (Noddings 2003)? If children love to move—an assumption here that I think many, including Ingold, would affirm (Gray 2013)—then is it better to teach self-control, discipline, stillness (Foucault 1977; Shlaes 2011)? Or is it better to allow random, free movement? Is culture to surrender entirely to nature?

IV

The nature(s) of humans is shaped by bipedalism, living in aggregated groups, imbued with sociality, connecting hand and brain, foot and brain, mapping. Some say that we are made by stories (Gottschall 2012), such as the Songlines mapping the Australian human-and-natural environment (Chatwin 1987), but always in motion. All these views suggest that as with growing plants, this nature may not really be violated.

So here in this volume we ruminate, roam-inate, on the nature(s) of humans, which surely must include walking. And those concerned with how to *e-ducare*, to educate, to lead out, must wonder, wander, about the emerging and emergent grasp of the incipient and never-finished process of becoming a human.

As Zhuangzi has it, "The Way Is Made by Walking."[1]

Note

1. There is a Chinese version, from Zhuangzi, of this phrase. In the twentieth century Lu Xun wrote that "the earth had no roads to begin with, but when many men pass one way, a road is made." The phrase is also attributed to the Spanish poet Antonio Machado, *se hace camino al andar*, and picked up by Paulo Freire and circulated in a rambling book he co-spoke with Myles Horton (Horton and Freire 1990; see also Freire [1970] 2000).

References

Benveniste, E. [1958] 1971. "Categories of Thought and Language." In *Problems in General Linguistics*, 55–64. Miami, FL: University of Miami Press.

Blum, S. D. 2013. "Cultivating Humans Sustainably: From Industrial Education to Permaculture. Transition and Solidarity for Sustainable Life on Earth." Paper presented at the 2013 Seoul Youth Creativity Summit and Festival, Seoul, South Korea.

———. 2016. *"I Love Learning; I Hate School": An Anthropology of College*. Ithaca, NY: Cornell University Press.

Bock, J. 2010. "An Evolutionary Perspective on Learning in Social, Cultural, and Ecological Context." In *The Anthropology of Learning in Childhood*, ed. D. F. Lancy, J. Bock, and S. Gaskins, 11–34. Walnut Creek, CA: AltaMira Press.

Chatwin, B. 1987. *The Songlines*. London: Jonathan Cape.

Chua, A. 2011. *Battle Hymn of the Tiger Mother*. New York: Penguin.

DeFrancis, J. 1984. *Chinese Language: Fact and Fantasy*. Honolulu: University of Hawai'i Press.

Dewey, J. 1966. *Democracy and Education: An Introduction to the Philosophy of Education*. New York: Free Press.

Edwards, C. P., K. Cline, L. Gandini, A. Giacomelli, D. Giovannini, and A. Galardini. 2013. "Books, Stories, and the Imagination at 'The Nursery Rhyme':

A Qualitative Case Study of the Learning Environment at an Italian Pre-school." In *Learning in and out of School: Education across the Globe*, ed. S. D. Blum. Conference proceedings. www.Kellogg.nd.edu/learning.

Elman, B. A. 2000. *A Cultural History of Civil Examinations in Late Imperial China*. Berkeley: University of California Press.

Fenollosa, E., and E. Pound. [1919] 2008. *The Chinese Written Character as a Medium for Poetry: A Critical Edition*. Ed. H. Saussy, J. Stalling, and L. Klein. New York: Fordham University Press.

Foucault, M. 1977. *Discipline and Punish: The Birth of the Prison*. Trans. A. Sheridan. Harmondsworth: Peregrine.

Freire, P. [1970] 2000. *Pedagogy of the Oppressed*. Trans. M. B. Ramos. New York: Continuum.

Gottschall, J. 2012. *The Storytelling Animal: How Stories Make Us Human*. Boston: Houghton Mifflin Harcourt.

Gray, P. 2013. *Free to Learn: Why Unleashing the Instinct to Play Will Make Our Children Happier, More Self-Reliant, and Better Students for Life*. New York: Basic Books.

Henrich, J., S. J. Heine, and A. Norenzayan. 2010. "The Weirdest People in the World?" *Behavioral and Brain Sciences* 33: 61–135.

Henry, J. 1963. *Culture against Man*. New York: Random House.

Holmgren, D. 2002. *Permaculture: Principles & Pathways beyond Sustainability*. Hepburn, Victoria (Australia): Holmgren Design Services.

Holt, J. [1976] 2004. *Instead of Education*. Boulder, CO: Sentient Publications.

Horton, M., and P. Freire. 1990. *We Make the Road by Walking: Conversations on Education and Social Change*. Ed. B. Bell, J. Gaventa, and J. Peters. Philadelphia: Temple University Press.

Illich, I. 1970. *Deschooling Society*. London: Marion Boyars.

Ingold, T. 2007. *Lines: A Brief History*. London and New York: Routledge.

———. 2014. "To Human Is a Verb." Paper presented at Human Nature(s) Conference, University of Notre Dame, April 4.

Kahn, C. H. 1966. "The Greek Verb 'To Be' and the Concept of Being." *Foundations of Language* 2 (3): 245–65.

Lancy, D. F. 2014. *The Anthropology of Childhood*. 2nd ed. Cambridge: Cambridge University Press.

Noddings, N. 2003. *Happiness and Education*. Cambridge: Cambridge University Press.

Rousseau, J.-J. [1762] 1979. *Emile, or On Education*. Trans. A. Bloom. New York: Basic Books.

Saussy, H. 2008. "Fenollosa Compounded: A Discrimination." In *The Chinese Written Character as a Medium for Poetry: A Critical Edition*, ed. H. Saussy, J. Stalling, and L. Klein, 1–40. New York: Fordham University Press.

Segall, M. H., D. T. Campbell, and M. J. Herskovits. 1966. *The Influence of Culture on Visual Perception*. Indianapolis, IN: Bobbs-Merrill.

Shlaes, L. 2011. "When a Child Can't Sit Still." *PediatricOT*, March 16. http://pediatricot.blogspot.com/2011/03/when-child-cant-sit-still.html.

Sivin, N., and G. E. R. Lloyd. 2002. *The Way and the Word: Science and Medicine in Early China and Greece*. New Haven, CT: Yale University Press.

Stevens, M. L. 2001. *Kingdom of Children: Culture and Controversy in the Homeschooling Movement*. Princeton, NJ: Princeton University Press.

Zhuangzi [Chuang-tzu]. 2003. *Zhuangzi: Basic Writings*. Trans. B. Watson. New York: Columbia University Press.

RESPONSE II

On Human Natures

Anthropological and Jewish Musings

RICHARD SOSIS

On Nouns and Verbs

My family loves word games. There is one game, however, that while always enjoyed nevertheless promises to provoke an argument. And it is always the same argument: what is a noun? Duple, as the game is known, requires players to form words that include the specific letters that they have picked from a central card pile. The type of word that must be formed is determined by another card pile, which indicates categories such as noun, verb, and adjective. After reading Tim Ingold's "'To Human' Is a Verb," I decided that during our next family game, when the appropriate letters were available, I would use "human" in the verb category. I tried, but when the opportunity arose, my children disallowed my move. Despite my protestations, including a detailed explanation of Ingold's argument, they came to a rare unanimous agreement: "human" is not a verb. I find that my children are generally a good litmus test for the

96

quality of my scholarly ideas—they are simply more grounded than I am—but I suspect the competitive atmosphere clouded their judgment in this case. I found much merit in Ingold's argument, although in the spirit of my role as commentator, I offer a few critical remarks with the aim of refining Ingold's astute thesis.

While Ingold's argument is novel, transforming nouns into verbs is an old academic pastime. For example, in attempting to rescue the increasingly ambiguous concept "mind" from the academic trash bin, the cultural ecologist Lesley White (1949) redefined mind as "minding." But as Geertz reminds us, nouns have never been just persons, places, and things.

> The use of nouns as dispositional terms—i.e., words denoting capacities and propensities rather than entities or activities—is actually a standard and indispensable practice in English, both natural and scientific. If "mind" is to go, "faith," "hope," and "charity" will have to go with it, as well as "cause," "force," and "gravitation" and "motive," "role," and "culture." (1973, 58)

No wonder all of our arguments while playing Duple involve the noun category. Geertz (1973, 58) continues that "mind" denotes a "class of skills, propensities, capacities, tendencies, [and] habits"; it refers to "an organized system of dispositions." Geertz's point is instructive. "Human," like "mind," is undoubtedly a dispositional term. Moreover, recognizing a thing as a process does not necessarily justify turning it into a verb, even if that thing becomes one with that process, as Ingold describes the human condition. The point is that Ingold's ideas about what it means to be human are interesting enough to stand on their own merit; creating a new language unnecessarily complicates matters. This is especially important in a discussion about human nature, which demands multiple disciplinary perspectives, each with its own disciplinary jargon that needs to be set aside in order to advance the conversation. To keep the lines of communication open, it would seem prudent to avoid neologisms, archaic terminology, and redefining colloquial terms. Maybe my kids had a point.

On Creating

One of the reasons I am sympathetic to Ingold's argument is that it is consistent with my own ethnographic and personal experiences in Judaism. But these same experiences also suggest a slightly different perspective than Ingold offers. During fieldwork in Tzfat, home of Jewish mysticism, I encountered an interesting Kabbalistic book titled *God Is a Verb* (Cooper 1997). Admittedly, I have difficulty penetrating mysticism, but the central idea of the book, revealed in its title, is straightforward. The etymological root of God's name in Hebrew is ambiguous, which has led to endless theorizing about its origin and meaning. About the only thing all of these theories agree on is that God's name is a verb. God is always creating.[1] I think the rabbis, therefore, would reject Ortega y Gasset's (1961) notion that God created the world in a single act, and finished the job, as Ingold puts it. Judaism insists that God never finished the job (Sacks 2005); humans are not re-creating, as Ingold, following Ortega, describes the human condition, but rather humans are co-creating. Indeed, if we are created in God's image, as Judaism maintains (*tzelem elokim*; Genesis 1:26), then Judaism apparently views humans as verbs as well.

On Walking

I was fascinated by Ingold's discussion of walking; Jan Masschelein's insight that walking pulls us away from any starting point is particularly intriguing. Here, however, I offer an alternative perspective. Jewish law, *halacha*, can be translated as "path" and is etymologically derived from the same root as the Hebrew verb "to walk." In Judaism, the goal of the walk is to reach messianic times (Neusner 1984), but the task—to fix the world on the way—is more important than reaching the goal. In the Talmudic tractate *Pirkei Avot* (2:21), for instance, Rabbi Tarfon asserts, "The task is not yours to complete, but neither are you free to desist from it."

Indeed, the rabbis strongly caution against taking messianic hopes too seriously. In Jewish law, for example, if you are planting seeds in your field and someone comes to tell you that the messiah has arrived, the rabbis instruct that you should finish planting and only afterward inquire about the validity of this claim (*Avot d'Rebbe Natan* 31b). Remarkably, this teaching is brought in the name of Rabbi Shimon bar Yochai, who Jewish tradition holds to be the author of the Zohar, Judaism's central Kabbalistic text. As commentators have pointed out, the task—planting—that Shimon bar Yochai highlights is telling. The individual is not just eating breakfast or cleaning the house; he or she is engaging in an activity that will not reap benefits for months until the crops are harvested. In other words, not only is the message of messianic arrival unlikely to be true today, but it is unlikely to be true in the near future as well. Therefore, the rabbis, including Judaism's preeminent kabbalist, advise: do not become distracted from tasks in this world, even mundane ones such as planting crops.

Jewish imagination, thus, entertains messianic hopes as a reality and a fabrication. While seemingly illogical and difficult to grasp, "contradiction," as the philosopher Ernst Cassirer (1944, 11) observes, "is the very element of human existence." He further opines that "religion cannot be clear and rational" (12), and therefore it is the only "approach to the secret of human nature[.] Religion is[,] . . . so to speak, a logic of absurdity; for only thus can it grasp the absurdity, the inner contradiction, the chimerical being of man" (11–12).

While "contradiction" may be at the heart of human existence, religion does not appear to be the only pathway to apprehension and resolution of our incongruous nature. In fact, we seem quite adept at living with contradiction. Consider our understanding of money. We realize that the green paper with presidential images has no inherent worth, but we would not tear it up because within the socially constructed world in which it operates, it has real effects and value. While logically inconsistent to hold an object to be both valuable and valueless, we seem to have no difficulty operating in this fabricated world (Searle 2010). The critical point here, and the revision I would like to suggest, concerns whether walking truly pulls us away from any standing point, away from the origin, as Ingold, following Masschelein, suggests. Jews do not always reflect on messianic goals—conceived as their original rationalization—when following halacha, but the messianic vision is with them as a resource in times of need. In other words, Jews carry the idea with them on their halachic journey, so to speak, and draw inspiration from it when such motivation is necessary. At the heart of Judaism is a tension, which recognizes that everything is temporary but that we need to engage with the world nonetheless. This appears to be accomplished by asking questions, yet never settling on answers beyond a specific time or place. The original questions continue to be carried on the walk, but one generation's answers are never fully satisfactory for the generations that follow. If the Jewish case is illustrative, then, as we travel many paths and unfold our lives we may not be pulled away from our origins to the extent that Ingold envisions.

A final thought about walking and human nature. Ingold will probably wish to resist this, but if the picture that he has described is accurate, and I believe it generally is, then it is not clear how we avoid the existence of a basic human nature that enables us to take the walk. Our cognitive and physical capacities allow us to take uncountable walks, all of which are unique at some level. But to claim that we cannot uncover some basic human nature—that is, the capacity for this metaphorical walking— might be confusing the phenomenological experience with the underlying capacity. It would seem that the latter is essential for the former to exist.

On Relationships

In his discussion of creativity, Ingold draws powerful insights from the American theologian Henry Nelson Wieman. Wieman usefully distinguishes between creativity that is "done" and creativity that is "undergone"; his suggestion that persons are created, or more specifically, undergo their creation, within community, is compelling. I agree entirely that "social life is not something that the person does but what the person undergoes: a process in which human beings do not create societies but, living socially, create themselves and one another." Wieman's position resonates strongly with Buber's (1970) observations on personhood and his notion that relationships define the human condition. The necessity of community for human vitality often seems lost in the fabric of contemporary societies, although religious communities, especially those that define themselves by what they do rather than what they believe, such as Islam and Judaism, insist that being human means standing in relation to the community (Tippett 2007).

However, it is unclear in Ingold's discussion of human life's "capacity to generate persons in relationships" whether he is describing a genuine human nature or rather an idealization that points to human potentiality. Jews, Muslims, and, of course, many other groups define persons in relation to their communities, but surely there are cultures, the contemporary United States being one strong candidate, in which the supremacy of individuality inhibits genuine relationships, thus stunting what Ingold and Wieman would label personhood. What are we to make of those who do not allow relationships to transform them into persons? Are they therefore not "humanifiers" or human? Is personhood an idealized goal, only achieved by some *Homo sapiens sapiens*? At the other extreme, if persons are only created in relationships, does this mean that some of us live multiple human lives? And if our lives are led by these relationships, as Ingold suggests, do each of us have as many lives as we have social webs?

On Fiction

As noted above, we are skilled at living in a fabricated world, but some of the fabricated worlds we create are meant to remain fabricated and not intended to be entertained as real. Ingold points us toward the past and the future, and I suggest we look at fictional pasts and futures to gain additional insights into human natures. Indeed, while many authors have explored human natures through interspecific comparative analyses (Rodseth et al. 1991; Wilson 2012), science fiction and fantasy offer almost limitless opportunities to comparatively examine other hominoids. In *Star Trek*, for example, every encounter with another life-form contains a tension in which the *Enterprise* crew is at pains to clarify what makes humans unique. Episode after episode, the captain and officers explain to "alien species" that humans differ from other life-forms by their need to grow. The entire mission of the *Enterprise*, to explore distant galaxies, is derived precisely from this need. And to the surprise of the many species they encounter on their journeys, for those on the *Enterprise*, life deprived of growth is not worth living; the denial of growth is death itself. Revealingly, the most dangerous nemesis of the United Federation of Planets is not the Romulans, Cardassians, or Klingons but rather the Borg, a cybernetic species who share the human desire for perfection but mistakenly believe they have achieved it, thus denying themselves the possibility for growth. In terms of our discussion, they've stopped walking on the path.

This key feature of human nature, growth, is not limited to a utopian future; fantasy writers have also explored this theme as rooted in our past. Consider Tolkien's writings. In Middle-earth, what distinguishes "the race of Men" from other "races" is specifically the need to grow (Dickerson 2012). Hobbits may "share a love of all things that grow,"[2] but Hobbits themselves do not change. They are anachronistically forever stuck in Tolkien's nineteenth-century English countryside. Orcs are interminably evil; repentance and redemption are never considered as an option for them. And the Elves, Ents, and Wizards of Middle-earth are depicted as eternal and unchanging, barely ever senescing. In Tolkien's mythology, lack of growth for the "race of Men" is inherently evil. Notably, the One Ring, the epitome of evil, not only prevents its bearer from physically aging, but it thwarts all aspects of human development and growth.[3]

It seems fitting, following a discussion of Middle-earth, to close with a reflection from the philosopher Paul Ricoeur (1988), who showed that fiction and history are less different than generally appreciated. Past events in human lives, what we often describe as history, did not unfold as a narrative. But to be human is to interpret the events of a life course as a story. If history, like fiction, is to offer meaning it demands a narrative structure. Therefore, while growth may be at the root of human natures, Ricoeur would caution us not to see our lives as progressing. Rather, our thirst for meaning, clearly one of our other human natures, necessitates that we interpret our lives, as individuals and communities, as developing. We not only need to grow; we need to know that we are growing.

ACKNOWLEDGMENTS: I thank Agustín Fuentes and Aku Visala for their invitation to participate in the Conference on Human Nature(s) and to contribute to this volume. I thank Jordan Kiper for very helpful comments on an earlier version of this commentary. Last, I thank Tim Ingold for an engaging and stimulating essay. This work was supported by the James Barnett Endowment for Humanistic Anthropology.

Notes

1. This conception of God is certainly not unique to Judaism. The influential medieval Muslim philosopher al-Ghazali, for example, conceives of God as movement (Aslan 2011).

2. Quoted from Peter Jackson's *The Fellowship of the Ring*.

3. *The Lego Movie* explores the same theme: ultimate evil is depicted as permanently gluing Lego worlds in place, preventing the possibility of change and growth.

References

Aslan, R. 2011. *No God but God: The Origins, Evolution, and Future of Islam*. New York: Random House.

Buber, M. 1970. *I and Thou*. New York: Charles Scribner's Sons.

Cassirer, E. 1944. *An Essay on Man*. New Haven, CT: Yale University Press.

Cooper, D. A. 1997. *God Is a Verb*. New York: Penguin Putnam.

Dickerson, M. 2012. *A Hobbit Journey*. Grand Rapids, MI: Brazos Press.

Geertz, C. 1973. *The Interpretation of Cultures: Selected Essays by Clifford Geertz*. New York: Basic Books.

Neusner, J. 1984. *Messiah in Context*. Philadelphia: Fortress Press.

Ricoeur, P. 1988. *Time and Narrative*. Vol. 3. Chicago: University of Chicago Press.

Rodseth, L., R. W. Wrangham, A. M. Harrigan, and B. B. Smuts. 1991. "The Human Community as a Primate Society." *Current Anthropology* 32 (3): 221–54.

Sacks, J. 2005. *To Heal a Fractured World: The Ethics of Responsibility*. New York: Schocken.

Searle, J. R. 2010. *Making the Social World: The Structure of Human Civilization*. Oxford: Oxford University Press.

Tippett, K. 2007. *Speaking of Faith*. New York: Viking.

White, L. A. 1949. *The Science of Culture: A Study of Man and Civilization*. New York: Farrar, Straus.

Wilson, E. O. 2012. *On Human Nature*. Cambridge, MA: Harvard University Press.

RESPONSE III

The Humanifying Adventure

A Response to Tim Ingold

Markus Mühling

Common Points of Departure

From a theological perspective, there are at least five decisive points in Tim Ingold's essay that should be regarded as a common basis of departure for interdisciplinary anthropological research.

1. Persons are relational entities. Human personality is shaped in reciprocally constitutive relationships to and from the other, be it with the personal other or with the so-called pre-personal realm.

2. Persons are becomings. The human person cannot be seen in essentialist categories but rather in an ontology of becoming. Humans do not possess narratives; they *are* narratives. Humans do not simply experience events; they *are* sequences or events in a broader framework of becoming. Humans do not simply experience development; they *are* developments.

3. The end of the becoming is not already set but open. Humans are not simply becomings, but there is a specific amount of contingency in the processes of becoming. The end is not programmed or fixed but, theologically speaking, promised. Humans are not epic but dramatic stories.[1]

4. Human personality is formed in a specific entanglement between passivity and activity, where the receptive capacity of perception is key. Or, in Ingold's words, "*Submission leads and mastery follows*: education as exposure precedes education as attunement."

5. Phenomenological approaches are necessary in order to prevent naturalism. Human persons are neither simple bodies, "extended phenotypes" (see Dawkins 1982, 82) of genes, nor souls possessing bodies, but they are *lived bodies*, a term that can be used as a translation for the German term *Leib*, in contrast to *Körper*.

Contributions of Theology

1. Persons, divine and human. Theologically speaking, all of the five features named above can be derived or at least combined with the vastly underappreciated definition of personhood by the twelfth-century theologian Richard of St. Victor. He defined a person as an *incommunicabilis existentia* (Richard von St. Victor 1958 4,18 [269]; and for an interpretation, see Mühling 2005, 161–64), a definition that most theologians (including Aquinas) failed to recognize in the right way. I'll therefore give a short explanation.

a) *Ex-sistentia* is one term combined from two: it does not mean being but signifies that any "being" consists in reciprocally constitutive (or internal) relations *from* and—as Richard says more or less explicitly— also *to* others. Therefore, the best translations for Richard's term *existentia* are "whence-and-whither-becoming" or "becoming-from-others-and-for-others."

b) *Incommunicabilis* also has two meanings. "Communication" in medieval times not only refers to verbal or nonverbal communication in a modern sense, but to any kind of transferal. It signifies, on the one hand, that there is something that is particular. In this sense one is nearly tempted to translate *incommunicabilis* as "individual." However, the second meaning is that despite the fact that the particularity cannot be communicated to others, it does not emerge prior to processual relationships to and from the others but *within* these communicative relationships themselves. This point is decisive since it safeguards the notion that human beings are communally constituted, especially in their particularity. The processual relationships to relata that are spatially external to the lived body of a person are constitutive for that person; that is, the spatially external relations are actually internal relations. This kind of communitarian thinking rejects both individualism and collectivism.

2. What is interesting is that Richard uses this definition of personhood in a univocal manner, first, for the trinitarian persons and, second, for created persons, in his case, humans and angels (see Richard von St. Victor 1958, 4,13–16 [256–64]). I would propose following Richard in these steps. Human persons have then to be seen not as "secondhand" gods, as in Ortega, but rather precisely with regard to those features Ingold has identified, traditionally called *imagines* (*sælæm* and *d^emût* [Heb.])—or as we might more aptly say, *resonances*—of God. As a result, God does not have to be seen as an exception to our ontological conception but rather as its primary instantiation and basis. This kind of thinking fits perfectly with the Hebrew tradition, which sees God as *'hjh 'ʃr 'hjh* (Exod. 3:14), as "I will be who I will be" or "I will be who will be with you." Neither translation is exclusive, and they can be combined. It is clear, however, that following these directions means that one has to look for a completely new ontology, which would be a relational ontology of dramatic narratives.[2]

3. For reasons I cannot go into here, it is decisive to see the features of humanity given by Ingold or the features of personhood given by Richard not simply as not distinguishing between the divine and the human but also as not distinguishing between humans and other creatures. There are surely a lot of humanifying features in other hominids, and who can say whether or not other vertebrates share in these? Under every circumstance we can also observe all decisive features mentioned above in the development of neonate humans (see Fuchs 2013, 188 f.). I would therefore like to suggest, in a heuristic manner, not speaking of the "human" or the "humanifying" as a traditional class, using traditional logic, but as

a fuzzy class, using fuzzy logic (see Zadeh 1965). The consequence would be that the "human" is dynamic and that there is a soft transition between the "human" and the "nonhuman." However, I do not think that the truth-function, which defines the fuzzy class of humanity, can be mathematically defined. Living personality, therefore, is a continuum in which any distinctions are possible but only for methodological purposes.

4. Theologically speaking, the humanifying development does not have a fixed end, nor is it completely open. The exposure of the displacement, which is inherent in the specifically human, has gone wrong and was not only dis-placed but also misplaced. The traditional notion for that development is *sin*. However, in the doctrine of the incarnation of God the Son and the doctrine of atonement it is said that Godself has personally carried out the misplaced "education" and opened new possibilities for a shared human and divine development. Therefore, the end is *promised* but not in a completely concrete form, which could then be treated as the kind of predetermined pertaining to the genre of epic. In the past I tried to express this distinction by differentiating between the ethical and the aesthetic shape of the ultimate or eschatical development: whereas the aesthetic shape of the ultimate reality is open and subject to contingency, the ethical outcome is guaranteed by the divine promise (Mühling 2004). The ideal perfect state of human being, therefore, is still becoming but a kind of becoming without ambiguities. However, even in this kind of perfect or eschatical becoming, exposedness to alterity has to be included (Mühling 2015, chs. 3 and 5). To use Ingold's words, the perfect state of becoming, called by Christians the perfection of the Kingdom of God, would be a nonclosed human (and divine) becoming or humaning in the attentional way of wayfaring without any intentional aspects of transport.

Tasks

Resonances Instead of External Relations

The question is not whether or not human becoming can be described relationally and processually but in what sense this processual relationality has to be understood. Here I think not only anthropology and theology but also the philosophical tradition of phenomenology, the postanalytical tradition of McDowell (2000) and the extended mind theory (Clark and Chalmers 1998; and recently Clark 2011), phenomenological approaches to the neurosciences (e.g., Fuchs 2013; Gallagher and Zahavi 2013), recent developments in the description of the theory of evolution such as niche construction and other developments on the way to an extended theory might be helpful (Odling-Smee, Laland, and Feldman 2003; for an overview, Fuentes 2009). Philosophically, it might be decisive to focus not so much on the pure external relationality of traditional efficient causality but to use concepts that stress internal relations even for seemingly external relata.

An extremely promising candidate for such an endeavor can be seen in the theory of the functional circuit, which can be traced back to Jacob von Uexküll (1973, 171; see also Fuchs 2013, 114 f.; Mühling 2014, 73–75)[3] and Victor von Weizsäcker and which replaces representational relations with the language of resonance. Whereas one can say that A might be a representation of B *but not* vice versa, one cannot say that A is a resonance of B *or* vice versa but only that A und B are resonating relata. These relata are not secondary or externally linked by relations but constituted by their relations itself, what can be called internal relationality between spatially external relata. In Ingold's terms, internal relata are not points secondarily connected by lines but the knots of meshed lines. There are many other questions here, but the points that Ingold has identified and that I regard as a point of departure mean that we are being exposed in our transdisciplinary endeavor to a joint journey—or to a shared adventure.

The Priority of Pathos

In a theological perspective, human becomings are creatures; that is, despite the truth that "to human" is a verb, this kind of activity has to be understood as a given or as a phenomenon in a strict sense. For example, Martin Luther saw that a completely *free will* is a *contradictio in adiectio* since the will cannot choose its own ability to make choices. Furthermore, it is dependent on bodily and affective states, meaning that will is always *bound will.*[4] Similarly, "to human" is also primarily a *pathos* that evokes a kind of responsivity (see Waldenfels 1997, 19). At this point the evolutionary embeddedness of humans and other creatures has to be slightly corrected. It may be the case that niche construction could provide the key component on the way to an extended theory of evolution. However, niche construction is by no means a kind of constructivism. For humans as well as for other organisms and populations there is no choice not to partake in niche construction. Strictly speaking, niche construction is always a response to *niche reception* (see Mühling 2014, 165). This kind of priority of *pathos* can be theologically understood as being created *sola gratia*. However, *pathos* in social and biological anthropology cannot simply be identified with being (or better, becoming) created in a theological sense. The attractive research task of looking for positive, neutral, and negative analogies emerges at this point.[5]

Loving the Other

The kind of permanent exposedness Ingold is stressing fits perfectly with the theological understanding of love as an ontological constitutive element of becoming human. In this sense, love is not motivated by deficits and needs but emerges due to the relatedness to the other that creates the longing. Love cannot be restricted to two relata, but it has to include a third one, which Richard of St. Victor calls the *condilectus* (1958, 3,11 [192]). There, love is not closed, but it creates openness, as Stanley Hauerwas stresses.[6] It would be a very interesting task to explore the possibilities of love as a concept of natural philosophy—a standard feature of medieval thinking (e.g., Dante Alighieri 33.143–45)[7]—in closer communication with Ingold's approach from social anthropology, with the biological concepts of pair bonds (e.g., Fuentes 2002) and the new role for cooperation in evolutionary theory (e.g., Fuentes 2004).

The Lived Body

If Ingold is right to see humans as becomings who are narratively exposed to others, they cannot be understood as either pure souls or pure bodies or as a combination of body and soul. Rather, as persons humans are always lived bodies; they have a *Leib* as a means for their communicative relatedness (Mühling 2015, ch. 5.2.4). From this perspective, the mind-body problem can only emerge from a wrong set of presuppositions. Phenomenologically, we experience the other and ourselves always as a lived body without any separation.[8] According to Husserl (1952, 63), we can adopt two attitudes, a naturalistic one and a personalistic one. However, both attitudes are primarily reflective abstractions from the prevalent observation of the lived body. Therefore, further research in bodiliness in the fields of biological anthropology, social anthropology, and theology—not only in the realm of theological anthropology, but also with regard to fundamental bodiliness in the doctrine of God—would be a fascinating task.

Conclusion

If we borrow these theological considerations and combine them with Ingold's fruitful considerations, I would like to give two more preliminary answers to the human question: To human is not only a verb, but humanifying or humaning also means not simply to undergo an adventure but also to *be* one. To borrow words from a poem by the no longer fashionable philosopher Ernst Bloch, I would like to offer as an answer to both the human question and our research on it:

> I am.
> But I do not have myself.
> Therefore, we become. (Bloch 1966, 13)[9]

Notes

1. Emphasis on the distinction between epic and dramatic narratives can be traced back to the theology of Hans Urs von Balthasar and was recently used, e.g., by Deane-Drummond (2009, 48–53).

2. For such an ontology, see Mühling 2014. Ingold has also discussed this theme; see, e.g., Ingold 2011, 141–76.

3. See Uexküll 1973, 171: "Every animal is a subject, which due to its specific constitution selects specific stimuli out of the general causes in the external world that it responds to in a specific way. These answers themselves have specific effects in the outer world, and in return influence the stimuli. This mechanism establishes a closed circuit that can be called the functional circle of the animal."

See also Fuchs 2013, 114 f.: "Every animal uses something like two grippers in order to be directed to its object—one organ of memory (receptor) and an organ of effect (effector). With these means it discovers the complementary qualities of reception and effection of the object, i.e., the animal gives the object the meaning of stimulus and cause."

4. I have given an extensive discussion on Luther's understanding of the will in discussion with neuroscientific research on will in Mühling 2013, 258–301.

5. The task of looking for positive, neutral, and negative analogies belongs to the theory of theoretical models presented in Hesse 1963.

6. Without knowledge about Richard's theology, Hauerwas has intuitively reformulated one of Richard's insights. Hauerwas 1983, 143 f.: "Thus the test of honesty of any relationship between two people is often the willingness of each to allow the other to begin a friendship with a third party that is not destined to be an equal friendship for both. For our friendships change us in ways we seldom anticipate, and when one of the other of us changes, our original relation changes."

7. See, e.g., Dante Alighieri, *Paradiso* 33.143–45: "But now my will and my desire, like wheels revolving with an even motion, were turning with the Love that moves the sun and all the other stars" (http://etcweb.princeton.edu/dante/pdp/). Love is foremost a cosmic principle that moves the inanimate world, and human becomings will only in their perfected state be able to join this principle.

8. This fact is also stressed in Ingold 1991.

9. "Ich bin. Aber ich habe mich nicht. Darum werden wir erst."

References

Bloch, E. 1966. *Tübinger Einleitung in die Philosophie I.* Frankfurt am Main: Suhrkamp.

Clark, A. 2011. *Supersizing the Mind: Embodiment, Action and Cognitive Extension.* Oxford: Oxford University Press.

Clark, A., and D. J. Chalmers. 1998. "The Extended Mind." *Analysis* 58: 10–23.

Dante Aligheri. *Divine Comedy.* Princeton Dante Project. http://etcweb.prince
ton.edu./dante/pdp/.

Dawkins, R. 1982. *The Extended Phenotype: The Gene as the Unit of Selection.*
Oxford: Oxford University Press.

Deane-Drummond, C. 2009. *Christ and Evolution: Wonder and Wisdom.* Minneapolis, MN: Fortress Press.

Fuchs, T. 2013. *Das Gehirn—ein Beziehungsorgan: Eine phänomenologisch-
ökologische Konzeption.* Stuttgart: Kohlhammer.

Fuentes, A. 2002. "Patterns and Trends in Primate Pair Bonds." *International
Journal of Primatology* 23 (5): 953–78.

———. 2004. "It's Not All Sex and Violence: Integrated Anthropology and the
Role of Cooperation and Social Complexity in Human Evolution." *American Anthropologist* 106 (4): 710–18.

———. 2009. "A New Synthesis: Resituating Approaches to the Evolution of
Human Behaviour." *Anthropology Today* 25 (3): 12–17.

Gallagher, S., and D. Zahavi. 2013. *The Phenomenological Mind.* London: Routledge.

Hauerwas, S. 1983. *The Peaceable Kingdom: A Primer in Christian Ethics.* Notre
Dame, IN: University of Notre Dame Press.

Hesse, M. B. 1963. *Models and Analogies in Science.* Notre Dame, IN: University
of Notre Dame Press.

Husserl, E. 1952. *Ideen zu einer reinen Phänomenologie und phänomenologischen
Philosophie 1. Husserliana 3/1.* The Hague: Nijhoff.

Ingold, T. 1991. "Becoming Persons: Consciousness and Sociality in Human
Evolution." *Cultural Dynamics* 4 (3): 355–78.

———. 2011. *Being Alive: Essays on Movement, Knowledge, and Description.* New
York: Routledge.

McDowell, J. 2000. *Mind and World: With a New Introduction.* Cambridge, MA:
Harvard University Press.

Mühling, M. 2004. "Why Does the Risen Christ Have Scars? Why God Did Not
Immediately Create the Eschaton: Goodness, Truth and Beauty." *International Journal of Systematic Theology* 6: 185–93.

———. 2005. *Gott ist Liebe: Studien zum Verständnis der Liebe als Modell des
trinitarischen Redens von Gott.* Marburg: Elwert.

———. 2013. *Liebesgeschichte Gott: Systematische Theologie im Konzept.* Göttingen: Vandenhoeck & Ruprecht.

———. 2014. *Resonances: Evolution, Neurobiology and Theology: Evolutionary
Niche Construction, the Ecological Brain and Relational Theology.* Göttingen:
Vandenhoeck & Ruprecht.

———. 2015. *The T&T Clark Handbook of Christian Eschatology.* New York:
Bloomsbury.

Odling-Smee, F. J., K. N. Laland, and M. W. Feldman. 2003. *Niche Construction: The Neglected Process in Evolution*. Princeton, NJ: Princeton University Press.

Richard of St.Victor. 1958. *De Trinitate*. Paris: J. Vrien.

Uexküll, J. 1973. *Theoretische Biologie*. Frankfurt am Main: Suhrkamp.

Waldenfels, B. 1997. *Topographie des Fremden I: Studien zur Phänomenologie des Fremden*. Frankfurt am Main: Suhrkamp.

Zadeh, L. A. 1965. "Fuzzy Sets." *Information and Control* 8: 338–53.

RESPONSE IV

The Ontogenesis of Human Moral Becoming

Darcia Narvaez

Ingold masterfully pulls our attention toward the nature of human nature. I would like to emphasize three points and add to Ingold's insights and review.

Humans as Biosocial Becomings

"Human beings do not create societies but, living socially, create themselves and one another," Ingold writes. Indeed, converging empirical evidence from across the neurobiological and developmental sciences indicates that humans are dynamic systems whose initial beginnings influence later capacities (Narvaez, Panksepp, et al. 2013; Narvaez, Valentino, et al. 2014). Early life sets the functional parameters and thresholds for many physiological systems (e.g., immune system, neurotransmitters, stress response). Each animal evolved a nest for its young, providing supportive care that matches up with the maturational schedule of the infant to optimize development (Gottlieb 1991). For social mammals, the nest evolved to fixation over 30 million years ago (Konner 2005). Human evolution intensified the needs of infants postnatally as they were born increasingly immature (9–18 months early compared to other animals) in order to make it through the birth canal that shrank to allow for biped-alism (Trevathan 2011).

The human nest or evolved developmental niche (EDN) for human youngsters includes mutually responsive relations with caregivers, exten-sive breastfeeding and positive touch, free play with multiaged mates, soothing perinatal experience, and positive social support. These charac-teristics provide a species-typical environment for a period when the brain and body are rapidly developing and establishing thresholds and param-eters for multiple systems. When characteristics of the nest are degraded or missing, the nest can be termed species-atypical, shifting the develop-mental trajectory to be species-atypical as well. My colleagues and I find that the characteristics of the EDN influence a child's self-regulation, con-science, and empathy, capacities that shape social life in childhood but also in adulthood. We find that missing EDN components lead to a sub-optimal pathway for sociality in children and in adults with not only worse health outcomes but also more self-focused moral orientations (Narvaez, Wang, et al. 2013; Narvaez, Gleason, et al. 2013; Narvaez, Wang, and Cheng 2015; Narvaez, Wang, et al. n.d.).

The reliance of humans on others for their becoming suggests that there may be more and less human ways to "become" from an evolution-ary systems perspective (Narvaez 2014). Since much of human adaptation has to do with sociality rather than things like muscular strength or acute vision, when early supportive care is not forthcoming, compromised self-regulation of multiple systems can compromise well-being and soci-ality over the long term, affecting one's personality and capacities. Over generations such undercare may weaken the germ line, influencing adap-tation (after all, one must outcompete one's rivals over generations to show fitness or adaptation; Lewontin 2010). Shepard (1998) points out that the whole maturation schedule of a human, from birth to adulthood, depends on a set of social supports that when missing leads to psycho-logical underdevelopment.

Receptive Intelligence

Part of what develops in the EDN is the right hemisphere, the seat of many self-regulatory but also receptive systems (Schore 2003a, 2003b). Although Ingold did not use the term *intelligence*, I am using it to denote what he describes as imagination. In the broadest sense imagination requires not only flexibility, but receptivity. Receptive intelligence is flexibly responsive in the moment rather than following a set of rules or script. As Ingold implies, humans evolved as movers within a natural world where receptive intelligence was vital. Such movement cultivates accurate perception and skills to live in and with one's local habitat. "Walking continually pulls us away from *any* standpoint . . . ," he writes. "It is a practice of *exposure*. And this is precisely what is achieved by education in the sense of *ex*-duction." Walking supports a receptive attention rather than the impositional, focused attention and emotional detachment of categorization.

Robert Wolff (2001) provides the richest descriptions of this approach to being. He describes the Senoi of Malaysia with whom he spent considerable time: they would "drift" during the day, going this way and that, without an explicit plan in mind. In Western terms they would have an intuition about something, like waiting by the road or going toward a particular tree, and the purpose would reveal itself once they followed the impulse to act (such as the appearance of a visitor or the discovery of a ripe fruit). Wolff writes, "Their existence had no reality until they lived it. They did not plan their lives they did not say to themselves or to each other, *Today we do this or that*" (2001, 120). Similarly Ingold states, "Thus the walker, a master of the terrain, must wait for signs that reveal the path ahead, with no surety of where it will lead."

Ingold, who has studied the ways of hunter-gatherers, emphasizes the type of learning and knowing through exploring the environment that characterized human learning prior to humanity's withdrawal into living mostly inside buildings: "He must watch his step, and listen and feel as well. He must, in a word, pay attention to things, and adjust his gait accordingly. Thus in the very act of walking, he is thrust into the presence of the real." Similarly, Wolff (2001) describes what he called learning to be human again. It consisted of walking through the jungle aimlessly with his "guide." After several visits of aimless days and increasing frustration about what he was supposed to be learning, Wolff says he decided just to be open to the life around him.

[I] stopped abruptly. The jungle was suddenly dense with sounds, smells, little puffs of air here and there. I became aware of things I

had largely ignored before. It was as if all this time I had been walking with dirty eyeglasses—and then someone washed them for me;
or as if I were watching a blurry home movie—and then someone
turned the focusing knob. But it was more than that—much more.
I could smell things I had no name for. I heard little sounds that
could be anything at all. (2001, 156)

Stepping into the "real," Wolff's perception was educated over time
as his guide patiently expected, much as Gibson's (1979) theory of direct
perception discerned. Immersion forms intuitions, building implicit understandings about the world (Hogarth 2000). Receptive intelligence
avoids the limitations of approaching the world as a set of objects but
takes experience as it comes, staying away from the emotional distancing
and detaching that categorization otherwise promotes.

The Importance of Embodiment and Embeddedness in the Natural World

Ingold and Wolff emphasize experience in the natural world, the place
where humans used to spend all their time. During the course of several
shifts in social organization over centuries, modern civilizations have dis-
tanced themselves from an integral relation to the natural entities around
them (Shepard 1998). This includes not only a withdrawal physically
from nature, but adoption of an emotional, oppositional relationship to
nature—humans against nature—often accompanied by disgust toward
nature. These attitudes are most clearly seen in the accounts of Europeans
encountering the "New World." They brought a cultural worldview that
made them blind to the paradisiacal and biodiverse landscape and fearful
of the "savage" wildness that they tried to control through intentional
destruction, human and other-than-human alike (Turner 1994). Negative
attitudes toward the natural world were of course contrary to those of
the many indigenous peoples Europeans encountered. These peoples had
no thought of moving against nature but sought to "maintain proper
relationships with [natural] beings" (Ingold 1999, 409). But even more
relevant here, they perceived life as an ongoing unfolding organic change,
a "continuous birth. . . . One is continually present as witness to [the]

moment, always moving like the crest of a wave, at which the world is about to disclose itself for what it is" (Ingold 2011, 69). This is a different form of being—rather like an ongoing *becoming*—from what is expected and fostered in modern civilizations.

One might ask what kind of biosocial being one becomes if one develops an us-against-them attitude toward nature. Elsewhere I have argued that the minimal capacities among many moderns to be relationally present to the nonhuman world and even to other humans are a result in part of early life trauma and undercare (lack of EDN), when capacities are initially established (Narvaez 2014). Early life undercare undermines synchrony with caregivers, the cornerstone of social life (Feldman 2014), and can lead to stress-reactive individuals who are unable to calm themselves except with emotional detachment, social domination, or endless consumption. These orientations lead to cultural practices that perpetuate the disordered ontogenetic cycle across generations.

Walking and living in natural landscapes where one can deeply learn the ways of the other-than-humans in one's neighborhood may be necessary for humans to fulfill their animal natures and develop accurate perception of their relationship to the biosphere. The lack of immersion in nature may have the opposite effect, undermining both perception and imagination. In fact, emotionally detached and nature-detached imagination seems to dominate assumptions in economics, business, and science. Relational responsibility toward others, human and other-than-human, is discarded as unscientific, bad business, and irrelevant to economic models (Latour 2013). Such detached imagination can undermine the health of both human and nonhuman creatures and the biodiverse planet that preserves us all. As Wolff (2001, 2–3) states, "By divorcing ourselves from Nature we have also removed ourselves from the wisdom that comes from living as part of What Is."

Human Becoming with (instead of against) Nature

Ingold has introduced us to the importance of a cultivated biosocial imagination, using walking as a demonstrable practice of receptivity. Our embodied lived experience guides us in developing accurate perception and capacities for perception, whether early social experience that builds social neural networks or early immersion in nature that shapes an alternative worldview and sensibilities from a human spending life inside walls.

Human becoming for most of the history of the genus meant living in nature with a deep sense of connectedness and receptivity to other-than-humans. And this requires immersed experience. To move away from the "wrong path" (Wolff 1999) of emotional and relational detachment, we must *move* in and with the natural world. As Evernden (1999, 77–78) points out, "So long as we have limited direct experience of the creature in question, it is relatively easy to accept a cultural stereotype in its place, in this case animal-as-object." We might say that many moderns have become autistic in relation to other-than-human nature—unable to hear, perceive, or understand their communications—unlike societies of the past (called animistic or panzooist; Halton 2014) who learned from the wisdom of other-than-human life.

To live well, one must have accurate perception. As Shepard (1998, 149) writes, "Perception's truest expression is its contiguity with nature, by which it influences the quality of life, our awareness of ecological integrity, and the connectedness of all things . . . [;] habit . . . that defines a group's pattern of bodily movement and sensibility, the predisposition emerging from genetic past and early grounding, affecting every aspect of one's expressive life." To have accurate perception, one must move through the habitat with capacities for and cultivation of receptive intelligence. For the social habitat, as described above, babies are best immersed in supportive early care, which facilitates the optimal shaping of their emotional and perceptive capacities (Turnbull 1983). For the natural habitat beyond humans, a similar immersion is required. If being in the natural world fosters a nature-attached becoming, then immersion there is required as well.

Ecology is therefore learning anew to-be-at-home in the region of our concern. This means that human homecoming is a matter of learning how to dwell intimately with that which resists our attempts to control, shape, manipulate, and exploit it (Grange 1977, 136). Now it is true that we cannot go back to living as nomadic foragers. But we can be more conscientious about ontogeny—how we raise children and integrate them into the natural landscape, preparing them for living cooperatively with nature. Ingold's extensive descriptions (across several publications, including Ingold 2013 and this vol.) shed light on the pathway back to wiser, more prudent and sustainable living.

References

Evernden, N. 1999. *The Natural Alien: Humankind and Environment.* 2nd ed. Toronto: University of Toronto Press.

Gibson, J. J. 1979. *The Ecological Approach to Visual Perception.* Boston: Houghton Mifflin.

Gottlieb, G. 1991. "Experiential Canalization of Behavioral Development: Theory." *Developmental Psychology* 27: 4–13.

Grange, J. 1977. "On the Way towards Foundational Ecology." *Soundings* 60 (1): 136.

Halton, E. 2014. *From the Axial Age to the Moral Revolution.* New York: Palgrave Macmillan.

Konner, M. 2005. "Hunter-Gatherer Infancy and Childhood: The !Kung and Others." In *Hunter-Gatherer Childhoods: Evolutionary, Developmental and Cultural Perspectives*, ed. B. Hewlett and M. Lamb, 19–64. New Brunswick, NJ: Transaction.

Ingold, T. 1999. "On the Social Relations of the Hunter-Gatherer Band." In *The Cambridge Encyclopedia of Hunters and Gatherers*, ed. R. B. Lee and R. Daly, 399–410. New York: Cambridge University Press.

———. 2011. *The Perception of the Environment: Essays on Livelihood, Dwelling and Skill.* London: Routledge.

———. 2013. "Prospect." In *Biosocial Becomings: Integrating Social and Biological Anthropology*, ed. T. Ingold and G. Palsson, 1–21. Cambridge: Cambridge University Press.

Latour, B. 2013. *Modes of Existence.* Cambridge, MA: Harvard University Press.

Lewontin, R. 2010. Response to Comment on "Not So Natural Selection." *New York Review of Books*, May 27.

Narvaez, D. 2014. *Neurobiology and the Development of Human Morality: Evolution, Culture, and Wisdom.* New York: W. W. Norton.

Narvaez, D., T. Gleason, L. Wang, J. Brooks, J. Lefever, A. Cheng, and Centers for the Prevention of Child Neglect. 2013. "The Evolved Development Niche: Longitudinal Effects of Caregiving Practices on Early Childhood Psychosocial Development." *Early Childhood Research Quarterly* 28 (4): 759–73. doi: 10.1016/j.ecresq.2013.07.003.

Narvaez, D., J. Panksepp, A. Schore, and T. Gleason, eds. 2013. *Evolution, Early Experience and Human Development: From Research to Practice and Policy.* New York: Oxford University Press.

Narvaez, D., K. Valentino, A. Fuentes, J. McKenna, and P. Gray. 2014. *Ancestral Landscapes in Human Evolution: Culture, Childrearing, and Social Wellbeing.* New York: Oxford University Press.

Narvaez, D., L. Wang, and A. Cheng. 2015. "Evolved Developmental Niche History: Relation to Adult Psychopathology and Morality." *Applied Developmental Science.* http://dx.doi.org/10.1080/10888691.2015.1128835.

Narvaez, D., L. Wang, A. Cheng, T. Gleason, and J. B. Lefever. n.d. "The Importance of Touch for Early Moral Development." Unpublished manuscript.

Narvaez, D., L.. Wang, T. Gleason, A. Cheng, J. Lefever, and L. Deng. 2013. "The Evolved Developmental Niche and Sociomoral Outcomes in Chinese Three-Year Olds." *European Journal of Developmental Psychology* 10 (2): 106–27.

Shepard, P. 1998. *Coming Home to the Pleistocene.* Ed. F. R. Shepard. Washington, DC: Shearwater/Island Press.

Trevathan, W. R. 2011. *Human Birth: An Evolutionary Perspective.* 2nd ed. New York: Aldine de Gruyter.

Turnbull, C. M. 1983. *The Human Cycle.* New York: Simon and Schuster.

Turner, F. 1994. *Beyond Geography: The Western Spirit against the Wilderness.* New Brunswick, NJ: Rutgers University Press.

Wolff, R. 2001. *Original Wisdom.* Rochester, VT: Inner Traditions.

CHAPTER 3

RECOGNIZING THE COMPLEXITY
OF PERSONHOOD

Complex Emergent Developmental Linguistic
Relational Neurophysiologicalism

WARREN BROWN AND BRAD D. STRAWN

In philosophical and theological anthropology (as in other fields), theoretical positions are typically expressed by short descriptive labels.[1] For
example, theories about the nature of persons with reference to embodiment are designated by labels such as body-mind dualism (or body-soul
dualism), emergent holistic dualism, emergent monism, dual aspect monism, and nonreductive physicalism—to name but a few. For the sake of
philosophical argument, positions and models are stated in conceptually
simple abstractions, involving telegraphic oversimplification. The problem becomes that these labels hide as much as they reveal, and allow discussion of their merit to proceed at an overly simplistic level in relation
to the topic at hand. It is our contention that, with respect to an understanding of human nature, overly general and abstract descriptors yield
to forms of argument that are at considerable distance from anything that
can adequately be about a functioning human being.

The danger to be avoided is falling into the sort of category error that comes about by choosing nomenclature that is too broad and abstract, and therefore discussions based on such categories hide the fundamental dynamic of what one intends to describe and characterize. Philosophical discourse that proceeds on the basis of large abstract categories that have no inherent anchoring in human life and functioning may be good philosophy, but it is not very relevant to the topic of human nature as it is lived and experienced. Of course, philosophical and theological anthropology cannot recapitulate all of neurophysiology. However, these domains of discourse must meet halfway what is known in the field of human neurophysiology by at least expressing models in ways that allow the inherent complexity and basic functional properties of hypercomplex neural systems like human brains and bodies to be explicitly recognized.

For example, we have argued in the past for "nonreductive physicalism." While much can be elaborated from this position, it is itself a minimalist philosophical statement about the fundamental physical nature of humankind but with the strong qualifier of having properties that are emergent and therefore not reducible to the properties of the elemental constituent parts. However, arguments for or against this position proceed on the basis of abstractions that, because of their simplicity, can systematically miss the point of the nature of the embodiment of personhood.

In this essay we propose a better descriptive title for a nonreductive physicalist position on human nature that, while wordy and cumbersome, makes explicit the complexity that needs to be encountered and that cannot be avoided in discussions regarding the nature of persons. We elaborate on each term in order to suggest what aspects of personhood to which it refers and the information that cannot be glossed over in discussions and debates.

Basic Positions on Human Nature

For most of its history, the Western Christian world has been dominated by the view that persons have a dual (two-part) nature: body and soul. This position (*dualism*) asserts that humans are composites of two different parts, a material body and a nonmaterial mind or soul (terms that are synonymous in most historical thought). In dualism, these two parts are not equals since the soul/mind is considered to be superior to the body and to rule over it. The soul/mind is the source of human rationality, sociality, and spirituality. It is also posited to be the locus of personal identity. In addition, the soul is immortal, whereas the body is mortal and transitory.

This sort of dualism is difficult to maintain in the light of modern neuroscience in that there is scarcely any human capacity that has not already been shown to be a product of identifiable patterns of brain activity. In addition, dualism raises the problem of how a nonmaterial soul would interact with a physical body. We have argued elsewhere that dualism is also problematic for reasons of its impact on human life, practical theology, and particularly communities (Brown and Strawn 2012). If the soul is superior to the body and rules over it, then life (one's own and that of others) must focus on caring for and nurturing the soul, first and foremost. One's body and outward behavior are secondary. Only if time and energy permit should attention be paid to the physical, economic, and social well-being of other persons.

In psychology, philosophy, and theology there are a number of alternative models for capturing the fundamental nature of human beings and personhood that are more resonant with neuroscience and neuropsychology. These models are expressed in terms such as those mentioned above: emergent holistic dualism, emergent monism, and so on. Very generally, these alternative positions take more seriously human neurobiology, as well as, among others, theories of dynamic systems and modern philosophy of mind. In most of these alternative formulations, the concept of neurobiological embodiment and emergence plays a central role.

Views of the embodiment of human personhood have been expanded in recent discussions in the philosophy of mind, as well as views that emerge from ongoing neuroscience research. Theories of *embodied cognition* and *extended mind* (Clark 1997) make it clear that the brain interacts dynamically with all of the peripheral systems that control and sense the entire body,[2] as well as extra-bodily tools and systems, in order to create the intelligent things persons do. In essence, we think with records of experiences gained as our bodies interact with the world, and we act more intelligently as we involve aspects of the external world. Human nature is emergent from more than just a complex brain, but also from entire bodily systems of motor control and sensory feedback.

While we embrace the basic motivation of these more holistic and embodied understandings of personhood, it is the conceptual simplicity of most of these that we wish to address in this essay. We propose a more complex, and therefore more robust, view, which we refer to using the cumbersome but richer name: complex emergent developmental linguistic relational neurophysiologicalism (CEDLRN, if you like acronyms). The point of labeling the position in this way is to make absolutely explicit the richness and complexity necessary for an adequate model of human nature. From this viewpoint, personhood is constituted by emergent properties that are the product of self-organizing processes within the hypercomplex neurophysiological systems of human beings and that come about progressively over a long period of developmental, linguistic, and relational history. To make this position clear a word-by-word clarification is necessary and provided in what follows.

Neurophysiologicalism

We turn first to the idea of neurophysiologicalism, since it is the noun that the other terms modify. Our reasoning for substituting "neurophysiological" for "physical" is that the category "physical" (or even "biological") is too general and abstract for a theory of human nature. "Neurophysiological," at minimum, refers to a living biological organism with a functioning nervous system—that is, complex networks of interactive neurons. At the very least, "neurophysiology" implies a high level of biological complexity (hypercomplex in the case of a human being) and an ongoing functioning system that is dynamically coupled with the world. Without a view that explicitly encompasses this sort of system, one misses critical points about human nature. For example, arguments about emergence in other sorts of physical systems (e.g., solid state physics) are grossly inadequate for capturing what is critical about the emerging properties that come about within functioning neurophysiological systems.

In addition, among other things, "neurophysiologicalism" implies *ongoing* dynamic feedback control and a hierarchy of levels of processing and control in moment-to-moment behavioral interactions with the world. Any attempt to conceptually halt this ongoing dynamic (as in philosophical arguments involving a "current brain state") has already missed the point of neurophysiological coupling with the environment via rapid feedback adjustments modulated by many layers of increasingly complex criteria of evaluation.[3]

Finally, we would be equally happy with the term *neuropsychologicalism* in that discussions of human nature are necessarily focused on the functional properties from which personhood emerges—that is, the neurophysiological processes that give rise to the psychological phenomena of thought, emotion, memory, planning, and religious experience. The question to be asked is whether there is merit in considering such critical properties of persons to be emergent from human brain-and-bodily neurophysiology, and for personhood to be further emergent from dynamic interactive patterns of such neuropsychological phenomena.

Complex

The attribute of complexity in our model is meant to reflect not only the fact that the human nervous system is arguably the most complex structure known but also the fact that the theory of complex dynamic systems provides a robust functional model for explaining how the high-level properties of the human mind emerge and play a causal role in unfolding human personhood.

First, the human brain is physically complex beyond what it is possible to imagine. The human nervous system is estimated to be composed of something like 100 billion neurons interacting over 100 trillion synapses. Recent work has demonstrated that just the presynaptic bouton terminal is composed of over 300,000 copies of 80 different molecules (Wilhelm et al. 2014). Despite the fact that potential patterns of neuron interactivity are constrained by the physical architecture of the brain and nervous system (what is connected to what), given the plasticity and possible degrees of functional connectivity over each synapse, the range of functional patterns that can be expressed from moment to moment by the human brain is truly astronomical.

Degrees and forms of structural connectivity are as important as sheer size in contributing to the unique cognitive capacities of humans.

Comparative neuroanatomy has made it clear that while humans do not have the largest brains, they have a relatively larger cerebral cortex and, most markedly, a very large prefrontal cortex. The area of the prefrontal cortex is roughly 4 percent of the total cerebral cortex in a cat, 10 percent in a dog, 12 percent in a macaque, and 17 percent in a chimpanzee; but in the human brain it has enlarged to 29 percent of the total cerebral cortex (Fuster 2002, 374–76). Enlargement of the human prefrontal cortex has been found to result primarily from an increase in the amount of *white matter*—that is, the degree of interconnectivity (Schoenemann, Sheehan, and Glotzer 2005, 242–52). Thus the human prefrontal cortex is not simply larger, but more intensely and complexly interconnected within itself, and with other cortical and subcortical structures, than that found in our primate cousins.

Because of the complex, highly interactive, and functionally plastic nature of the human nervous system, the *theory of complex dynamic systems* provides a robust model for understanding the impact of complexity and interactivity. When an aggregate of interactive agents (e.g., neurons, ants, groups of people) gets pushed far from equilibrium by environmental pressures, a complex dynamic system will often emerge as the aggregate organizes itself into *patterns* of interactivity. Such patterns are constituted by *relational facilitation and constraint* between elements such that they come to work together in a coherent or coordinated manner to create a larger-scale functional system that can adapt to the demands of the physical, social, or cultural environment in complex and subtle ways. Thus *aggregates* of disorganized elements become *systems* when the probability of each element doing one thing or another is altered, constrained, and entrained by interactions with other elements, and the new properties emerge based on system organization.

The physical requirements that are necessary for the self-organizational processes that give rise to a dynamic system and the emergence of higher-order properties include *complexity* (a very large number of elements, such as neurons), a high degree of *interconnectivity* (e.g., axons, dendrites, and synapses), two-way interactions (recurrent connections, feedback loops), and *nonlinear interactions* that amplify small perturbations and small differences in initial conditions. Thus the complex and massively interconnected neuronal network that is the human brain, and most particularly the cerebral cortex, is perfectly suited for the emergence of high-level causal properties-of-the-whole through dynamic self-organization.

Emergent

As noted above, dynamic systems theory gives the best account of how emergence takes place.[4] In this account, emergence denotes the fact that complex interactive *systems* can have properties that do not exist within the *elements* that make up the system but emerge from complex patterns of interactivity. Once organized into a functional system, the lower-level properties of the elements interact (bottom-up) with the relational constraints created by the higher-level patterns of the system (top-down). Continuing interactions with novel aspects of the environment cause repeated reorganizations of these interactive patterns that allow immediate adaptation and at the same time build into the system increasingly more complex and high-level repertoires of possible responses to future challenges from the environment.

The fact that systems are the product of self-organization with respect to interactions with, and feedback from, the environment means that the system organization embodies *meaning* that is carried forward in the form of memory of previous interactions and the relevant system reorganizations. By action and consequent feedback, a system learns to deal with an immediate novel environmental challenge, and also becomes prepared to deal meaningfully with similar situations in the future.

In this manner, the establishment of patterns of *constraint* between lower-level elements results in the emergence of higher-level properties that manifest *greater freedom* in the form of a larger response repertoire. The system has a substantially greater number of possibilities with respect to its interactions with the surrounding environment than it had prior to each new level of self-reorganization. "By enlarging and coordinating previously aggregated parts into a more complex, differentiated, systematic whole, contextual constraints [between elements] enlarge the variety of states the system as a whole can access" (Young and Saver 2001, 78).[5] Thus whole repertoires of interactional possibilities emerge over time from the processes of self-organization and the ongoing environmental challenges that trigger reorganization.

Thus dynamic systems theory specifies how truly emergent, non-reductive properties can be manifested by complex, interactive, near-chaotic, and contextually embedded physical systems—most particularly within the hypercomplex brains of human beings. Mental processes (e.g., thinking, deciding, consciousness, memory, language, representation, belief) are self-organizing dynamic patterns that create top-down influences on the lower-level neurophysiological phenomena whose activity constitutes the mental processes themselves. In the end, the emergent properties of complex dynamic systems transcend the descriptions of their constituent parts, demanding a different level of descriptive language to do justice to the nature and activities of a system as a whole.

Developmental

How do we become the complex interactive persons that we are? During fetal and infant development, cells of the brain and spinal cord are forming into neurons, finding their place in the vast complexity of brain space, hooking up with neighboring neurons, and sending axons out to distant regions of the brain. During this process, the outcome is only very roughly preordained by the genome. Emerging neural *structure* and interconnectivity is maximally open to *functional* organization as the physiology of the developing human responds to its internal milieu and to interactions with the external environment. The chaotic physical fumbling and amorphous interactions of a newborn infant ultimately result in progressive adaptation to the physical, social, and cultural world that the child inhabits.

Thus the functional structure of the brain is formed ("wired") through a kaleidoscope of daily experiences. Genes contribute only a rough blueprint of brain wiring; the rest is formed by a self-organizing process based on continued feedback from interactions with the environment. The development of intelligence, personality, and character, while influenced to some degree by genes, mostly takes place through action and feedback. The human mind is a combination of certain very general predispositions, and the experience-based emergence of mental capacities, personality, and character through a continuous history of situational and social interactions. Exploration, give-and-take, and trial and error with the physical and social environment fundamentally change us at the level of our neurons, even as adults.

This experienced-based, self-organizing nature of human mind and personhood is significantly enhanced in humankind by the very slow physical development of the cerebral cortex (Quartz and Sejnowski 2002). Whereas a chimpanzee has a nearly adultlike cortex (for a chimp) by the end of the second year of life, a similar degree of development is not reached in humans until four to five years later. In fact, the cortex is still maturing in the late teenage years. This slow development is particularly characteristic of the frontal lobes, areas critical for the most sophisticated processes of the mind. Prolonged physical development is important for allowing maximal opportunity for brain wiring to be sculpted by the physical, social, and cultural world of the individual. Thus differences in mental power and sociality between humans and apes are not simply due to differences in the size of the brain, but also to the significantly extended opportunity for social and cultural learning to influence the organization of its fine structure.

The point we wish to emphasize is that intelligence, personality, and character are mostly developmentally "open programs" in the same way that some computer super-games have been created to learn game knowledge as they are played. The program itself gets progressively modified and improved by the experiences of trial-and-success and trial-and-error feedback as the game is played; so also with the developing brain of a child. Thus our neural openness gives us a great advantage in mental flexibility, allowing our personhood to be shaped for unanticipated roles and challenges. Theories of human nature that do not explicitly take into account the dynamics of child development typically fail to take into account a primary source of much of what is uniquely human.

Linguistic

In terms of the emergence of human nature and qualities of personhood, the development of language is a game changer. Sometime during the second year of life, language learning kicks into high gear and accelerates exponentially. With language comes new cognitive, social, and psychological capacities that fundamentally alter personhood.

The capacity for language seems to have both a genetically related predisposition and a critical developmental (self-organizing) aspect. The existence of an early development blueprint for language has been suggested in studies using functional magnetic resonance imaging (fMRI) of brain activity. When prelinguistic infants listen to speech, fMRI has shown evidence of localization of activity in left hemisphere language areas. However, it is also clear that language is impossible for children to learn outside of rich and constant language exposure from parents and other persons. Thus mastery of language is a process of self-organization in response to a language-dominant social environmental. For example, infants start life with the ability to tell the difference between all human language sounds (phonemes), but over the first few years of life they lose the ability to detect differences between sounds that do not occur in their native tongue (Kuhl et al. 1992, 606–8). If the mother tongue they are learning does not include a particular phoneme distinction (e.g., *ra* vs. *la*), they soon are no longer able to "hear" the difference. In fact, if language is not heard and used during childhood due to some form of social or sensory deprivation, the language capacity can suffer permanent loss. The linguist Susan Curtis (1997) has documented just such an outcome in the socially isolated case of Genie.

As the capacity for language develops it provides the basis for the emergence of important mental and social capacities, some of which have been outlined by the anthropologist Terrence Deacon (1997). First, language distances behavior from the demands of immediate motivations and needs. Language can be used as a means of postponing gratification of immediate needs by considering alternative actions using language-dependent thoughts about the future. Second, the self-referential possibilities of language allow a person to have thoughts and express sentences that include the idea of "me." In this way, language facilitates the formation of a self-concept, allowing self-reflection, self-understanding, and projection of the self into the future. Third, language provides the basis for expanded empathy and an ability to understand the mental lives of others. We enter into, and become emotionally engaged by, the experiences of others through stories. Fourth, language facilitates a common mind among groups of people. Common semantics, metaphors, and stories form groups with similar understandings of life and the world around them. Children are brought into the worldview of their parents, family, and, eventually, wider social groups via language. And fifth, ethical behavior is enhanced by language because it holds and communicates the values of communities in terms of statements and stories about what is "good" and "bad."

Storytelling is particularly important in the emergence of personhood because it allows persons to know vicariously the experiences of others. It is particularly important in child development. A story narrative allows children to vicariously imagine new situations, "try out" various behaviors, and safely experience these behaviors' positive or negative consequences, forming rich impressions in their minds about what is good and bad, right and wrong, and conducive or not to the well-being of others. Narrative memories also contribute to our understanding of our selves. This is illustrated in a rare neurological condition called *dysnarrativia* that is caused by damage to a particular part of the frontal lobe of the brain. Persons with this disorder are impaired in their ability to formulate a narrative history of their thoughts and experiences, even though they are quite aware at the moment of their circumstances and thoughts. Individuals with this disorder "lead 'denarrated' lives, aware but failing to organize experience in an action-generating temporal frame" (Young and Saver 2001, 78). Most interestingly, these individuals are described as having lost their sense of self, suggesting a strong link between the capacity to narrate one's experiences and important qualities of personhood.

Relational

As we have explained, it is impossible to formulate an adequate understanding of human nature and personhood without taking explicit account of the role of developmental processes in forming personhood and the contribution of language to human cognitive and social capacities. It is equally impossible to formulate an understanding of human nature and personhood without an account of the fundamental relationality of humans. While some philosophical positions on anthropology give some recognition to human relationality, they seldom fully appreciate the power of human interpersonal interactions in the development of personhood and the depth of its impact on what it means to be a person. The role of relatedness cannot be ignored or minimized in an adequate theory of human nature.

The prolonged developmental openness of the human nervous system is obviously not sufficient to ensure the formation of personhood in a child. The openness of neural organization is coupled with innate behavioral tendencies to interact with other persons. As the psychologist Andrew Meltzoff has famously shown, even within hours of birth infants will imitate the facial gestures of another person. Imitation of facial expressions begins a social give-and-take between parent and child (Meltzoff 2007, 26–43; Meltzoff and Decety 2003, 491–500). Such tendencies to engage in reciprocal imitation build into the cognitive system of a child the idea that others are "like me," an important step in the organization of an understanding of the mind of other persons (called theory of mind). Thus, based on innate primitive social inclinations and a developmental process of self-organization, children form social relationships, which serve to elicit and deeply form their personhood.

In addition to an inclination to interact with other persons, psychologists have shown that human infants are born with a particular temperament, the most extreme types being "inhibited" (20 percent of infants) and "bold/fearless" (40 percent).[6] However, by four years of age, only 10 percent of children show such extremes in temperament. Thus the experiences of life during these early years tend to modify these temperaments. In the self-organizing process of human development, being born "*some* way" does not equal being forever "*that* way" (Quartz and Sejnowski 2002, 128). Even more striking is research that bred monkeys to be primarily inhibited (fearful) or bold in temperament (presumably a genetically mediated selection process). Later, some of the fearful/inhibited infant monkeys were moved into a cage with uninhibited, nurturing foster mothers. They subsequently became less fearful, as indicated by a reduction in biochemical markers of stress and fear (Champoux et al. 2002, 1058–63). Relational experience significantly modified a genetically influenced temperament.

Extensive research has explored the impact of early attachment experiences on the subsequent development of a person. Different qualities of the early caregiver-child relationship lead to core forms of self-organization that constitute the child's developing sense of self-in-the-world, self-and-other, and emotional self-regulation. These core forms come together to create an identifiable style of adult relationships, referred to as an *attach-ment style* (Cozolino 2002; Siegal 1999). Four basic attachment styles have been identified in children: securely attached, anxiously attached and avoidant, anxiously attached and resistant, and disorganized attachment (Ainsworth et al. 1978; Main and Solomon 1986, 95–124). Secure attachment doesn't just magically happen but is accomplished through a large number of child-caregiver interactions that bring about what has been called neural organization and integration (Cozolino 2002).

As we would expect from our understanding of the possibilities for reorganization in dynamic systems (such as the human brain), there is hope for children who did not experience a secure base from which to de-velop a secure attachment. Research has demonstrated that attachment styles can change if given new relational experiences over time.[7] While longitudinal research has indicated that attachment styles (i.e., character-istics of early relational organization) typically endure into adulthood, it has also shown that good friends, caring spouses, and good psychotherapy can ameliorate negative attachment styles and move individuals toward more secure styles (i.e., relationally provoked system reorganization).

Tragically, the importance of early childhood caregiver relational interactions has been demonstrated in extreme situations of deprivation (Curtis 1977; Spitz 1946, 53–74). "Genie" (a pseudonym) was thirteen years old when she was discovered by Child Protective Services in a home in Los Angeles. Since very early childhood she had been subject to severe social isolation and deprivation. She had been strapped to a potty chair during the day and caged in a crib all night. She was never spoken to, and there were few ambient sounds of conversation, television, or radio to be heard outside her room. When first discovered, Genie was thought to be autistic and had no language capacity. After extensive language training, she eventually acquired some vocabulary, but she was never able to de-velop normal phonology or more than minimal competence in grammar. Of course, her social interactions were severely abnormal, with particular problems with anger management and childlike behavioral interactions. It took her several months just to learn how to smile. As an adult, she lapsed into a state similar to dementia (Curtis 1977).

Genie was autistic-like, socially inept, linguistically deficient, and generally mentally disabled, not because of some brain disorder. Rather Genie was the victim of severe childhood social deprivation. She had lived for thirteen years without the usual love of parents, interaction with other people, and stimulation gained from exploration of the physical and cultural world. She is a tragic illustration of the fact that, despite all the genetic influences that result in normal human bodies and brains, persons do not come preformed. Rather at birth we are mentally and socially unformed, amorphous, plastic, and open to being shaped by the environment.

Unfortunately, Genie is not the only child who has suffered from such social neglect. Revelation of the condition of children living in Romanian orphanages after the demise of the dictator Nicolae Ceaușescu was shocking. Because of the economic and social policies of Ceaușescu, thousands of parents had no other choice but to leave their children to be raised by state-run orphanages. The mass housing and feeding in these institutions was markedly devoid of human contact, care, and love. While basic physical care was provided, there was a very minimal amount of any other form of human physical or verbal interaction. Much like Genie, some of these orphaned children developed symptoms reminiscent of autism (although they were not autistic). They would sit in their dirty cribs and rock themselves endlessly. In addition, the mental development of these orphans was impaired (Nelson et al. 2007, 1937–40). Interestingly, recent research has shown that if these children were placed into the homes of loving foster families within the first two years of life, the negative impact of these dire conditions on their mental development was substantially reduced.

Conclusion

We began this essay by arguing that most of the models currently framing the conversation about human nature suffer from two problems: (1) there is an implicit inner-outer dualism inherent in many models that otherwise claim to be monist; and (2) all of the current models and theories of human nature express their positions in terms that are too sparse and therefore fail to bring into the conversation an array of dimensions of human nature that must be explicitly understood and taken into account before progress can be made. This essay has focused on the second issue by putting forward a more explicit and complex model of human nature: complex emergent developmental linguistic relational neurophysiologicalism.

We have argued that we must not be content to merely philosophize about physicalism without being cognizant of the unique characteristics of neurophysiology. What is more, the human brain is the most complex structure and system known. This complexity must be explicitly acknowledged and understood as having the properties of a complex dynamic system, that is, subject to processes of self-organization from which emerge high-level, causal, agentive neurocognitive properties. Adequate understanding of human nature must also take into account the progressive process of self-organizing neurocognitive development that is characteristic of human children, as well as the dramatic contribution to psychosocial capacities and human agency of the acquisition of language. Finally, the deeply formative influences of interpersonal relatedness and sociocultural systems must be taken into account.

It is our position that progress cannot be made in theories of human nature without this sort of explicit recognition of the depth and richness of what it means to be a human being and to become a person. Given the state of understanding of complex systems, human cognition, neuroscience, child developmental psychology, and relational psychology, it is time to move the discussion of human nature past the sparse models currently employed.

Notes

1. Some of the topics discussed here are also discussed in a chapter in Joshua Farris and Charles Taliaferro, eds., *Ashgate Research Companion to Theological Anthropology* (Aldershot: Ashgate, 2015).

2. See, e.g., Teske 2013.

3. These ideas are developed more completely in Jeeves and Brown 2009.

4. See Juarrero 1999.

5. Ibid., 138.

6. Research by Jerome Kagan described in Quartz and Sejnowski 2002, 125.

7. Research described in Wallin 2007.

References

Ainsworth, M. D. S., M. C. Blehar, E. Waters, and S. Wall. 1978. *Patterns of Attachment: A Psychological Study of the Strange Situation.* Mahwah, NJ: Lawrence Erlbaum.

Brown, W., and B. Strawn. 2012. *The Physical Nature of Christian Life: Neuroscience, Psychology and the Church.* Cambridge: Cambridge University Press.

Champoux, M., et al. 2002. "Serotonin Transporter Gene Polymorphism, Differential Early Rearing, and Behavior in Rhesus Monkey Neonates." *Molecular Psychiatry* 7: 1058–63.

Clark, A. 1997. *Being There: Putting Brain, Body, and World Together Again.* Cambridge, MA: MIT Press.

Cozolino, L. J. 2002. *The Neuroscience of Psychotherapy: Building and Rebuilding the Human Brain.* New York: W. W. Norton.

Curtis, S. 1977. *Genie: A Psycholinguistic Study of a Modern-Day "Wild Child."* New York: Academic Press.

Deacon, T. 1997. *The Symbolic Species: The Co-Evolution of Language and the Brain.* New York: W. W. Norton.

Fuster, J. M. 2002. "Frontal Lobe and Cognitive Development." *Journal of Neurocytology* 31: 374–76.

Jeeves, M. A., and W. S. Brown. 2009. *Neuroscience, Psychology, and Religion: Illusions, Delusions, and Realities about Human Nature.* Radnor, PA: Templeton Press.

Juarrero, A. 1999. *Dynamics in Action: Intentional Behavior as a Complex System.* Cambridge, MA: Bradford Books, 1999.

Kuhl, P. K., et al. 1992. "Linguistic Experience Alters Phonetic Perception in Infants by 6 Months of Age." *Science* 31: 606–8.

Main, M., and J. Solomon. 1986. "Discovery of an Insecure/Disorganized Atttachment Pattern." In *Affective Development in Infancy*, ed. T. B. Brazelton and M. Yogman, 95–124. Norwood, NJ: Ablex.

Meltzoff, A. 2007. "The 'Like Me' Framework for Recognizing and Becoming an Intentional Agent." *Acta Psychologica* 124: 26–43.

Meltzoff, A., and J. Decety. 2003. "What Imitation Tells Us about Social Cognition: A Rapprochement between Developmental Psychology and Cognitive Neuroscience." *Phil. Trans. Royal Society of London* B 358: 491–500.

Nelson III, C. A., et al. 2007. "Cognitive Recovery In Socially Deprived Young Children: The Bucharest Early Intervention Project." *Science* 318, no. 5858 (December): 1937–40.

Quartz, S., and T. Sejnowski. 2002. *Liars, Lovers, and Heroes: What the New Brain Science Reveals about How We Become Who We Are*. San Francisco: Harper-Collins.

Schoenemann, P. T., M. J. Sheehan, and L. D. Glotzer. 2005. "Prefrontal White Matter Volume Is Disproportionately Larger in Humans than in Other Primates." *Nature: Neuroscience* 8 (2): 242–52.

Siegal, D. J. 1999. *The Developing Mind: How Relationships and the Brain Interact to Shape Who We Are*. New York: Guilford Press.

Spitz, R. 1946. "Hospitalism: An Inquiry into the Genesis of Psychiatric Conditions in Early Childhood." *Psychoanalytic Study of the Child* 1: 53–74.

Teske, J. A. 2013. "From Embodied to Extended Cognition." *Zygon* 48: 759–87.

Wallin, David J. 2007. *Attachment in Psychotherapy*. New York: Guilford Press.

Wilhelm, B. G., S. Mandad, S. Trukenbroldt, et al. 2014. "Composition of Isolated Synaptic Boutons Reveals the Amounts of Vesicle Trafficking Proteins." *Science* 344 (6187): 1023–28.

Young, K., and J. L. Saver. 2001. "The Neurology of Narrative." *Substance: A Review of Theory & Literary Criticism* 30: 78.

RESPONSE I

"Self-Organizing Personhood" and Many Loose Ends

LLUIS OVIEDO

Developing models of human nature is not an easy task. It is easy to fall into reductive positions, which are unable to account for complex processes, or to get lost amid a broad set of variables of unpredictable outline. With great skill Warren Brown and Brad Strawn avoid both dangers and supply an accurate model of human nature and its development. This new model integrates several features in a pattern of "emergence" that suggests a self-organizing, complex, and dynamic system that describes how humans are constituted and behave. Since this model pays close attention to neurological explanations, it is invested with scientific rigor and updated insight.

Thus far the proposed model is much more satisfying than former ones, which were content with showing the unavoidable physical dimensions of human nature and trying to overcome explicit or latent expressions of dualism. Brown and Strawn's account is extremely important for building a more realistic anthropology and having interdisciplinary strength. "Christian anthropology" should pay close attention in order to update its content.

Surely the present project converges with other attempts to integrate distinct dimensions of human nature, most notably the one proposed by Jablonka and Lamb (2005), which has been followed by others (e.g., Fuentes 2009). Unfortunately this pattern suggesting a more complex representation of the human has not been extensively applied in related fields concerning human activities or some of its more specific features, such as art, language, morality, or religion. A look at the main literature in the field of the new scientific study of religion provides a good example. Indeed, those trying to better account for religious cognition and experience during the past fifteen years have followed simplified and reductive models of human mind and behavior, mostly based on computational schemas and straightforward adaptive rules (Barrett 2010).

Anthropological models have many implications, not always rendered explicit in the current research on human faculties or behaviors. Traditionally, it has stated the close dependency between the study of human beings and the study of religion. Such affinities and mutual implications are quite apparent when looking at modern philosophy and the many attempts to come to terms with both our representation of human nature and our understanding of religion. Recent developments in the scientific study of religion often imply hidden or undeclared anthropologies (Oviedo 2012). Here I review the proposed model and compare it to recent developments in the study of consciousness, language, and religion in order to assess its meaning and limits.

In my opinion, two main concerns arise from Brown and Strawn's brilliant construction: the first has to do with the place reserved for consciousness and its functions; the second, with the application of the concept of emergence. In several cases, the narrative of their brief essay reveals almost automatic processes, which gradually configure the main traits of human nature and give rise to a very complex being, emerging from lower properties. Even the semantics, "self-organization" and "emergence," sound quasi-mechanical, as if human nature works without conscious means that would include decisions, free will, and the exercise of discernment and self-control. One important issue in the proposal under examination is whether the described process of anthropological emergence is more or less contingent, or rather necessary, and the result of natural forces moving toward the present constitution of personal subjects.

The main point, suggested by part of the title of this essay, "Self-Organizing Personhood," is that personal identity is the result of emerging instances that give rise to the actual configuration of human beings. In that emergent schema, neurological structure is completed by "linguistic" and "relational" emergent properties, which explain the current complexity of human beings.

Several points in the supplied schema raise the question of whether such a pattern would work without a means of control in place, a "self," in the sense of a personal self-conscious subject deciding what is more convenient in each case. Taking the most abstract level, dynamic complex systems have been described in different ways. One of them, the theory of social systems, formulated by the German social theorist Niklas Luhmann, states that some systems proceed through "meaning," that is, selections among various possibilities that have to be steadily reduced to find the right course of action (Luhmann 1981). Personal systems do this function primarily through conscious decisions, even if that "reduction of complexity" works in many other ways, and "meaning" may become just a kind of "bias" or heuristics helping one to decide without too much reflection.

Consciousness and *meaning* are elusive terms with several meanings and referents. To be sure, at least twice the authors refer to "meaning" and "meaning making" in the context of emerging properties and child development; nevertheless, something is left out when conscious aspects are ignored or considered just as emergent properties. A more holistic or integrated anthropology should recognize them and give them a role in the description of human nature.

A large amount of research and theoretical reflection has tried in the past decade to make sense of the conscious, reflective dimensions of human mind and behavior, to challenge the most reductive and "eliminativist" positions, which would "discharge" conscious process from any significant role (Baumeister, Masicampo, and Vohs 2011). Very roughly, these accounts characterize conscious, reflective, and controlled cognition as a necessary complement to more intuitive, elementary, and automatic forms. This point has been highlighted by the so-called dual process theories (Kahneman 2011; Evans 2008; Stanovich 2011). Such an approach shows a complex system in human cognition in which at least two complementary strategies interact to provide better, faster, and more effective ways to address different challenges in human life. Many of them require a simple, fast, and automatic answer, to save time and effort; many others demand a lot of attention, analysis, and "slow thinking." Kahneman argues for introducing reflective and critical means to tackle bias—that is, to *de-bias*—bound to intuitive cognition. The problem is that often such biases are a source of confusion and wrong perceptions, with fatal consequences. This is not to suggest, however, that the reflective mode is superior to the intuitive and unconscious one but that it is a form of "complementarity," since the intuitive mode is very useful in many settings as well. In any case, what emerges here is a model of cognitive complexity, a "dual system" that can hardly be reduced to a simpler mode of reasoning.

Furthermore, in recent years studies have shown the presence of an entangled set of first-order and second-order emotions interacting with cognition and influencing human behavior (Izard 2011). The main point is that emotions can be elaborated through cognitive and cultural means, to produce personal positive or negative states. Beyond the pattern of a more mechanical or biological framework, which could explain most of human behavior, the idea of "elaborated emotions" entails a more active involvement of human minds and conscious rumination. Again meaning arises in such interaction, since the choices one makes are strongly guided by one's emotional reactions and cognitive elaborations. So once more complexity seems unavoidable.

Similar consideration needs to be given to the embodied and embedded character of human nature, which cannot be isolated from social interaction and language. Again, this process accompanies—as Brown and Strawn show—the emergence of a mind that is able to interact and of a language that provides new cognitive skills. However, even if this process would have an almost mechanical character, once in place it would work in a less predictable way, opening itself up to infinite possibilities, regulated mostly by human consciousness. An entire line of research reveals the dynamics of cultural influence in human cognition and behavior interacting in a fruitful way with other dimensions of human cognition (Tomasello 1999).

To test the model thus far provided by Brown and Strawn, the cases of language and religion could help assess the proposed emergent framework. The problem consists in part of giving more substantial content to what is understood by "emergence." This term could mean many things and requires a more focused explanation of how humans and their "special features" arose and evolved. The case of language is paradigmatic. A recent article signed by eight prominent specialists in linguistics, neurology, and evolutionary studies states, "Based on the current state of evidence, we submit that the most fundamental questions about the origins and evolution of our linguistic capacity remain as mysterious as ever" (Hauser et al. 2014, 1). The article reviews the evidence provided by four main lines of research, showing their limits or their inability to provide a convincing picture of the origins—or emergence—of human language.

Then there is the case of religion. There is a clear parallel between attempts to deal with human nature from a scientific perspective and the new scientific study of religion. In the latter case, after a long season of reductive approaches, many voices now argue for the inclusion of cultural dimensions (Van Slyke 2011), looking at religion as a "meaning system" (Paloutzian and Park 2005), and for considering more reflective—and not only intuitive—forms of religious cognition, in what is generally called a multilevel approach (Whitehouse 2013). Paloutzian and Park state that meaning systems "comprise mental processes that function together to enable a person (religious or not) to live consciously and nonconsciously with a sense of relative continuity, evaluate incoming information relative to whatever his or her guidelines may be, and regulate beliefs, affects, and actions accordingly" (2005, 6–7). What we are learning is that any attempt to deal with religion without considering the conscious human mind is doomed to failure, as it is unable to account for a large array of religious phenomena.

Many theories have been proposed to explain in sheer naturalistic terms the origins and evolution of religion, or, in other words, its "emergence" from earlier or more basic features. In a systematic review still in progress I have gathered more than sixty studies by different authors. In many cases the theories can be clustered in big groups having several shared points, like those stressing the adaptive character of religion, or its cognitive structure, or its ritual patterns. However, my guess is that something very similar to what has been shown by Hauser and his colleagues concerning the mystery of the origin of language could be replicated regarding the origins of religion: all these theories are based on scarce evidence and have difficulties dealing with this very human feature—religion—involving so many factors and probably very closely linked in its origins and evolution to that of human language. What is coming to light in the former cases is that an "emergentist" explanation of some human features does not help overcome or suppress the mystery surrounding them. This problem has been seen in the search to explain consciousness, another human trait that is emergent but not yet explained in a satisfactory way. Many authoritative voices have claimed that we still lack a convincing theory based on empirical evidence that can provide a sufficient explanation of this faculty and its applications (Nagel 2012).

Now, summing up these virtual tests, the idea of "self-organizing personhood" should take into account the difficulties of giving a satisfactory account of the origins and evolution of such very human traits. In that case the "self-organizing hypothesis" has to be recognized as a possible explanation while trying to be truthful to the scientific ideal, that is, to be naturalistic, without falling into reductive and implausible descriptions of human nature. However, this is where we move beyond purely scientific analysis: questions need to be raised about the "directionality of the evolutionary process" itself, especially when observing the process that brought humans about. This is where a strictly naturalistic stance might become problematic.

This is also the point where dualism enters through the back door. In my opinion, if the need for conscious dimensions is acknowledged, and hence free will has to exist in decisional terms—beyond the developmental processes leading to mature humans—then we cannot avoid the issue of dualism, except that we pretend to ignore the reach of conscious, reflective forms, involved in meaning and purpose, and defining what we are and do. After all, conscious reflection is what renders us open beings, increasing in an exponential way the possibilities of development and interaction.

Anthropology now more than ever before has to confront a dilemma. On the one hand, it could provide an accurate description of human nature and its dynamism but end up with not so much a person as a robot. On the other hand, it could highlight more specific human traits, like free will and consciousness, but the outcome might be a too complex being, hardly fitting within a strictly scientific framework. Sometimes we may need to choose.

In the end, the real complexity of human nature has to do with the irreducible character of its conscious dimension, as has been argued in several recent essays (see, e.g., Nagel 2012). What renders humans really complex and unpredictable is their capacity to make decisions among many courses of action, taking into account rational, emotional, heuristic, cultural, and whatever other factors that cannot be described in algorithmic terms.

References

Barrett, N. F. 2010. "Toward an Alternative Evolutionary Theory of Religion: Looking Past Computational Evolutionary Psychology to a Wider Field of Possibilities." *Journal of the American Academy of Religion* 78 (3): 583–621.

Baumeister, R. F., E. J. Masicampo, and K. D. Vohs. 2011. "Do Conscious Thoughts Cause Behavior?" *Annual Review of Psychology* 62: 331–61.

Evans, J. St. B. T. 2008. "Dual-Processing Accounts of Reasoning, Judgment, and Social Cognition." *Annual Review of Psychology* 59: 255–78.

Fuentes, A. 2009. *Evolution of Human Behavior.* Oxford: Oxford University Press.

Hauser M. D., C. Yang, R. C. Berwick, I. Tattersall, M. J. Ryan, J. Watumull, N. Chomsky, and R. C. Lewontin. 2014. "The Mystery of Language Evolution." *Frontiers in Psychology.* doi:10.3389/fpsyg.2014.00401.

Izard, C. E. 2011. "Cognition Interactions—Forms and Functions of Emotions: Matters of Emotion." *Emotion Review* 3–4: 371–78.

Jablonka, E., and M. Lamb. 2005. *Evolution in Four Dimensions: Genetic, Epigenetic, Behavioral, and Symbolic Variation in the History of Life.* Cambridge, MA: MIT Press.

Kahneman, D. 2011. *Thinking, Fast and Slow.* New York: Farrar, Straus and Giroux.

Luhmann, N. 1981. *Soziale Systeme: Grundriss einer allgemeine Theorie.* Frankfurt am Main: Suhrkamp.

Nagel, T. 2012. *Mind and Cosmos: Why the Materialist Neo-Darwinian Conception of Nature Is Almost Certainly False.* Oxford: Oxford University Press.

Oviedo, L. 2012. "Do We Need to Naturalize Religion?" In *Is Religion Natural?*, ed. D. Evers, A. Jackelen, M. Fuller, and T. Smaedes, 85–102. London: Continuum Press.

Paloutzian, R. F., and C. L. Park. 2005. "Recent Progress and Core Issues in the Science of the Psychology of Religion and Spirituality." In *Handbook of the*

Psychology of Religion and Spirituality, ed. R. F. Paloutzian and C. L. Park, 3–22. New York: Guilford Press.

Stanovich, K. E. 2011. *Rationality and the Reflective Mind.* Oxford: Oxford University Press.

Tomasello, M. 1999. *The Cultural Origins of Human Cognition.* Cambridge, MA: Harvard University Press.

Van Slyke, J. A. 2011. *The Cognitive Science of Religion.* Farnham: Ashgate.

Whitehouse, H. 2013. "Rethinking Proximate Causation and Development in Religious Evolution." In *Cultural Evolution: Society, Technology, Language, and Religion*, ed. P. J. Richerson and M. H. Christiansen, 349–63. Cambridge, MA: MIT Press.

RESPONSE II

A Last Hurrah for Dualism?

KELLY JAMES CLARK

I would like to focus on just two paragraphs in the first section of Brown and Strawn's essay, where the authors make a number of claims about human nature. The remainder of their essay is a complex and learned defense of a kind of nonreductive materialism (which may actually be a kind of emergent property dualism). Along the way, I defend mind-body dualism, not because I am a dualist, but because I think it is an intellectually viable option that merits discussion (I am agnostic about the relationship of mind to body—with a slight inclination toward materialism about persons). But since our topic here is human nature, I decided to comment mostly on Brown and Strawn's opening remarks concerning human nature.

> For most of its history, the Western Christian world has been domi-
> nated by the view that persons have a dual (two-part) nature: body
> and soul. This position (*dualism*) asserts that humans are composites
> of two different parts, a material body and a nonmaterial mind or
> soul (terms that are synonymous in most historical thought). In dual-
> ism, these two parts are not equals since the soul/mind is considered
> to be superior to the body and to rule over it. . . .

This sort of dualism is difficult to maintain in the light of modern neuroscience in that there is scarcely any human capacity that has not already been shown to be products of identifiable patterns of brain activity. In addition, dualism raises the problem of how a nonmaterial soul would interact with a physical body. We have argued elsewhere that dualism is also problematic for reasons of its impact on human life, practical theology, and particularly communities. . . . If the soul is superior to the body and rules over it, then life (one's own and that of others) must focus on caring for and nurturing the soul, first and foremost. One's body and outward behavior are secondary. Only if time and energy permit should attention be paid to the physical, economic, and social well-being of other persons.

I want to address three of their claims, all of which are related to their construal of mind-body dualism but not all of which are developed in the above passage: (1) the claim that the mental cannot causally interact with the physical, (2) the claim that dualism has deleterious practical consequences, and (3) whether or not we really understand the concept of emergence.

The Mental and the Physical

How can a nonmaterial soul interact with a physical body? Can mind affect matter? We know that if we throw a rock through a window, the window will break. But just thinking about it will not break the window. We know that sticks and stones can break bones, but thinking it will not. So it is hard to imagine how, as Brown and Strawn put it, the mental could interact with, cause, the physical. Conceded.

But in the Christian tradition, there is a model for mind's ability to causally interact with matter. God created the world out of nothing and sustains it in existence. If that view is correct, Mind not only moves matter; Mind creates it out of nothing. Mind creates and shapes matter, first into earth and sky and then into creatures that inhabit earth and sky. Maybe God's creating the universe out of nothing is incoherent. It may even be metaphysically impossible for the mental to affect the physical. But there is a kind of internal inconsistency for Brown. He is a Christian who believes that God created the heavens and the earth. Thus he has a paradigmatic case for the mental causing the physical. It would be curious indeed for a Christian to offer the impossibility of the mental causing the physical as an argument against dualism. This ad hominem argument is not an argument for dualism, but it does show that if one believes that Spirit can create and sustain a universe, then one should likewise concede the possibility of a nonmaterial soul interacting with a physical body.

One might likewise think, to extend the Christian reference, that creation in God's image implies mind-body dualism. God, according to Christians, is spirit, and Christians likewise believe, as Brown and Strawn put it, that the human mind/spirit is "the source of human rationality, sociality, and spirituality." It is a small step from such considerations to mind-body dualism. Moreover, if God is eternal and immutable, yet bodies are "mortal and transitory," then creation in God's image suggests a soul/spirit that is immortal. Human beings, the stuff of clay and gods, are both body and mind. While I do not endorse this inference, it is a historically influential one.

The Christian tradition affirms and biblical writers assume mind-body dualism (Cooper 2000). But perhaps the biblical writers affirmed mind-body dualism not because they were divinely inspired in this regard but simply because humans are cognitively constituted to affirm or assume mind-body dualism (Bering 2006). Their writings include mind-body dualism because they were incapable of conceiving of humans in other ways. If this is the case, mind-body dualism may be on par with geocentrism: while humans are cognitively constituted to folk dualism and folk physics *and* folk dualism and folk physics are widely assumed by the biblical writers, belief in dualism would not be theologically required of Christians. After all, and in support of Brown and Strawn, Christians confess to the resurrection of the *body*, not the immortality of the soul.

And yet, though it may not be required of Christians, dualism still has some life left in it. Regarding the nature of human persons, substance dualism holds that both an immaterial mind and a material body exist as distinct substances; it is perhaps the most commonsensical way of understanding the nature of humankind. The psychologist Paul Bloom (2004), for example, argues that belief in dualism is innate in all humans. In support of dualism is the stubborn resistance of mind being reduced to the brain (i.e., explained fully in terms of chemical or neuronal processes). The philosopher Colin McGinn puts it this way: "The more we know of the brain, the less it looks like a device for creating consciousness: it's just a big collection of biological cells and a blur of electrical activity—all machine and no ghost."[1] Pain sensations, for example, are first-person experiences, which are not captured by third-person scientific accounts. While portions of the brain are active (an electroencephalogram might record the firings of neurons in your hypothalamus) and a neuroscientist might be able to identify the chemical processes involved, neither the brain activity nor the chemical processes are the first-person, felt experience of pain. C-fiber firings are not the pain, and chemical processes are not the pain. Pain is a feeling that is qualitatively distinct from the physical processes associated with it. Moreover, physical properties are qualitatively different from mental properties.

Brown and Strawn are partly right: scientists have demonstrated many *correlations* between the mental and the physical. And it would be silly to deny the mind's dependence on the brain. Yet there is not a single successful reduction of the feeling of pain to brain processes. The mental is qualitatively different from the physical. Perhaps the mind is not reducible to the brain.

There are other considerations that favor mind-body dualism. If the physical is relatively deterministic, then it seems to leave no place for free will (free will is, prima facie, difficult to square with determinism). And if there is no free will, then there seems to be no moral responsibility (moral responsibility, prima facie, assumes that people are free with respect to their choices). However, if there is a part of a person (a soul, say) that floats free of the physical, then it is possible that a person is not always subject to the physical laws of cause and effect. A soul, then, comports well with human freedom and its concomitant support for moral responsibility.

And a soul that can separate from the body comports well with afterlife beliefs. If one's body remains in, say, a grave, then issues of personal identity seem to favor one's soul (that which makes you essentially and uniquely you) separating from the body at death and moving on into the afterlife. If *you* survive death but your *body* does not, then you are not your body. One's soul could be reunited with a body, no problem there. Thus postmortem survival and identity favor mind-body dualism.

The subjectivity of first-person experiences (and their resistance to reduction to the physical) along with free will, moral responsibility and postmortem existence favor mind-body dualism. Brown and Strawn have offered some good reasons in favor of materialism. I do not think any of these cases is compelling (at this point of human inquiry). There is no decisive reason—biblical, experiential, philosophical, or scientific—that favors materialism over dualism or vice versa. While Brown and Strawn may hope that science will demonstrate materialism, prescience is not science. While a dualist may believe that the mental can cause the physical, she has not yet offered an account of how this is possible. On either view, we are left with a deep mystery, something important that is fundamentally unexplained. In the case of the materialist, we still have no idea how the mental could arise from the physical. In the case of the dualist, we still have no idea how the mental causes the physical. When mystery is compared to mystery, there is no decisive winner.

Dualism's Practical Consequences

I'm a philosopher, and I've grown increasingly aware of how irrelevant we are. We seldom discuss issues that have any impact on anything in the world whatsoever. Philosophy has grown increasingly abstract and disconnected from practical life. If Brown and Strawn are right, though, philosophy isn't really irrelevant. In fact, the metaphysics of persons has a huge impact on human life. If one is a mind-body dualist, they claim, one should not care for other persons' physical needs.

I'm dubious of this claim for several reasons. First, I think each of us, including a mind-body dualist, is psychologically disposed to care for at least our own bodies. We care for our bodies; we sleep and eat and bathe. And we spend an enormous amount of time thinking about, for example, our next meal (or our last meal). Not only do we care about our own bodies, but we are psychologically disposed to care about our childrens' bodies. We care that our own children are fed and bathed and get enough sleep. Perhaps we care inordinately in this regard for our own children's physical needs. The more difficult issue is our relative lack of care for the physical needs of nongenetic conspecifics. While I think we are relatively unmoved by the physical needs of nongenetic conspecifics, I think we are nearly equally unmoved by the state of their souls.

We have a good explanation of our concern for our own and our children's physical needs and our lack of concern for nonkin physical needs. And this explanation has no recourse to metaphysics. Evolution, so it seems, has shaped us to care inordinately for the vehicles carrying our genes—my body and my children's bodies—and very little for vehicles that carry others' genes. This scientific explanation of, for lack of a better term, selfishness, requires no appeal to the metaphysics of persons. And I think it, or something in the neighborhood, is true.

Evolution may have likewise shaped us to believe in souls. That is a different question. But evolution's outfitting us to believe in souls is not nearly as powerful as evolution's outfitting us to care for our own and our offspring's physical needs.

Finally, I think one's metaphysics of persons is morally benign. Whatever one thinks of the metaphysics of persons, one is morally obligated to care for people's physical needs because caring for people's physical needs is right and good. Christians, again to appeal to Brown and Strawn's religious allegiance, are required to care for the physical needs of others because God has commanded them to. If one has moral duties to care for other people's physical needs, one's metaphysical beliefs are thereby morally irrelevant.

My concluding point, and this (at last!) takes us off the paragraphs quoted above, concerns the concept of emergence. I think there is a lot of hand waving in discussions of the concept of emergence. We know that water emerges from hydrogen and oxygen and that water has different properties from its constituents, hydrogen and oxygen. And we know that clouds emerge from configurations of water droplets and that clouds have properties that individual water droplets do not. And yet we would be hard pressed to say exactly what emergence is. If you are a materialist about persons, you might say that free will is an emergent property; and, unless you are a reductive materialist about consciousness, you will think consciousness is emergent. So you will *think* that consciousness, for example, is emergent, but we do not really know how consciousness emerges from the matter that constitutes it. We do not really have an argument or even an understanding of emergence yet (for water, for clouds, or for consciousness). Calling it emergence makes it sound like we have explained it, but we really haven't. Unless or until we have a clear understanding of emergence, we do not understand how consciousness emerges from matter.

These comments should not detract from a fine essay, of which I have ignored nearly every important part. But those parts that impinge on issues of human nature are highly suggestive and worth pausing on for reflection. Brown and Strawn have raised their voices in this very important conversation, and their claims merit critical reflection.

Note

1. www.newstatesman.com/ideas/2012/02/consciousness-mind-brain.

References

Bering, J. M. 2006. "The Folk Psychology of Souls." *Behavioral and Brain Sciences* 29: 453–98.

Cooper, J. 2000. *Body, Soul and Life Everlasting: Biblical Anthropology and the Monism-Dualism Debate*. Grand Rapids, MI: Eerdmans.

RESPONSE III

Why the Foundational Question about Human Nature Is Open and Empirical

CARL GILLETT

Philosophers have long debated what I term the "Foundational Question" about human nature: What deeper kind of thing are we? Or, put another way, what kind of individual is a human? Recently, scientific and wider debates over human nature have flared to life. One might expect these discussions to engage the Foundational Question, but it is rarely explicitly addressed. The reason is that it is implicitly assumed that the sciences have long ago answered the question. Anthropologists, primatologists, evolutionary biologists, and many beyond the sciences following them all routinely assume that we are identical to *Homo sapiens* animals and that evolution established this "fact." What philosophers aptly term "Animalism" is therefore the position implicitly assumed about the Foundational Question in wider discussions. For instance, recent debates about nature versus nurture concern whether the animal we are is determined by environment or its innate nature, while discussions of our origins concern which primate species (*Homo habilis,* etc.) are the ancestors of our kind of animal, and so on.

155

I contend that the received wisdom is badly mistaken, though not for the types of reasons my fellow philosophers often press. For I believe it is recent scientific advances that show that the Foundational Question about our deeper nature is empirically resolvable but still open. At the outset, I should immediately note that I take it to be overwhelmingly plausible that we are evolved, since we are complex biological entities and evolution is the best explanation of the genesis of such entities. But if not through a challenge to evolution, where does the challenge to the received wisdom that we are identical to animals come from? Just as the earlier scientific revolution in biology driven by evolutionary frameworks shook the prevailing picture of our natures, so in our own time I contend it is the so-called neuroscience revolution that is remaking the intellectual terrain. Though the resulting issues are complex, my goal here is to provide a sense of why it is becoming increasingly plausible that the Foundational Question is open and we have yet to resolve the kind of individual we are.

The Neuroscience Revolution and Rich Psychological Properties

As the massive media coverage makes us all aware, the neurosciences are making rapid advances at a number of neurobiological levels. The brain is made of literally billions of neurons, 86 billion by one recent estimate (Azevedo et al. 2009), with still greater numbers of connections between them. And vast amounts of resources are presently being directed to brain mapping projects focused on understanding these connections and hence the so-called connectome. But we have already made profound advances at higher neurobiological levels where we now have increasingly detailed accounts of what I term "rich psychological" properties like remembering, fearing, or hoping.

For example, in cognitive neuroscience the work of Endel Tulving (2002) and others has provided us with sophisticated accounts of rich psychological properties like episodically remembering, that is, remembering a particular episode. In affective neuroscience, Jaak Panksepp (1997; Panksepp and Biven 2012) has given us accounts of properties like fearing. As these researchers emphasize, such accounts are based on multiple lines of evidence, including testing on human subjects, cases of lesions or damage in humans, evidence from animal studies, evidence from the molecular and cellular levels, and findings from brain imaging.

This work on our rich psychological properties has provided us with well-confirmed hypotheses that illuminate what philosophers refer to as the "roles" of such properties. Thus episodic memories are individuated by being produced by sensory organs, like eyes, producing other rich psychological properties, and producing muscle stimulations, usually in combination with other rich psychological properties. Given the nature of these characteristic productive roles, we can provide plausible arguments about which kind of individual instantiates remembering and fearing.

A wide range of individuals associated with us can be considered candidates for having rich psychology, but let me stick to the most obvious suspects. So let me assume the candidate individuals are a *Homo sapiens* animal or brain. And let me also note that scientific notions of parthood plausibly imply that parts and wholes do not productively interact with each other. However, eyes and muscles are parts of animals, so the *Homo sapiens* organism does not productively interact with its own eyes or muscles. Thus the animal does not play the right productive role to instantiate rich psychological properties because they are plausibly had by an individual that does productively interact with eyes or muscles. In contrast, the brain does productively interact with sensory organs and muscles, and does in similar ways fill the role of remembering or fearing, since our physiology is constructed to facilitate just such interactions. Consequently, our new neuroscientific accounts of rich psychology plausibly support rich psychological properties being instantiated in the brains that are the individuals that have the property of episodically remembering, fearing, and so on.

In fact, this is just what working scientists, like Tulving, Panksepp, and many others, now increasingly assume, since they routinely ascribe rich psychological properties to brains. Scientists have thus also moved from describing properties such as remembering as a "psychological" property to terming it a "neurocognitive" property. Wittgensteinian philosophers (Bennett and Hacker 2003; Pardo and Patterson 2013) increasingly attack such scientific practices of ascribing rich psychology to brains as committing a "mereological fallacy" of ascribing properties of wholes to their parts. Such philosophical critics assume scientists have missed our ordinary practices of ascribing rich psychology to animals rather than their parts, including any organ like the brain. However, it is actually the critics who are missing the detailed theoretical arguments of the type I sketched above that justify the scientific practices. Neuroscientists are not falling into a fallacy but are instead committed to a rich, explanatorily successful set of hypotheses supported by multiple lines of evidence. It is this theoretical framework that underwrites the scientific practice of ascribing rich psychological properties to brains rather than animals—and that hence also challenges the everyday practices, of ascribing rich psychology to animals, that the Wittgensteinians prioritize without any similar theoretical defense.

We Are Expansive Brains

I contend that these developments in the neurosciences necessitate our adopting a new account in the foundations of neuroscience that I term "Expansive Materialism" since the view frames the point that the neurosciences have recently *expanded* our understanding of the properties of brains to include rich psychological properties like remembering and fearing. Crucially, Expansive Materialism takes the sciences to posit rich psychology but at the highest neurobiological level since it accepts what we might term "expansive" (or "thinking") brains that instantiate rich psychology. Expansive Materialism thus contrasts with both of the opposing views that presently dominate the foundations of neuroscience, each of which was thrown up in responses to earlier developments in the sciences.

On one side, inspired by the neuroscience of the 1970s and 1980s, focused on the molecular and cellular levels, we have the Eliminative Materialism of writers like Paul Churchland (1981) and Patricia Churchland (1986), which contends that rich psychological properties will have no place in the sciences. However, cognitive and affective neuroscience now routinely posits rich psychological properties at the highest neurobiological level, so we should favor Expansive Materialism over Eliminativism. On the other side of the current debates, driven by the emergence of cognitive science, in the 1960s, 1970s, and 1980s, we have what I term the

"Separatist Materialism" attributed to Jerry Fodor (1968, 1975) and others that contends that the sciences should study rich psychological properties but claims that rich psychological properties are instantiated in a level of individuals *above* the brain. In this case, Expansive Materialism should be preferred over Separatism given our new evidence that rich psychology is plausibly instantiated in the brain.

The latter points illustrate some of the deep changes in the intellectual landscape wrought by the neuroscience revolution. And these implications plausibly extend to our understanding of our own deeper nature and hence to the Foundational Question. Neuroscientific evidence is all too often associated in the media with nihilist positions that claim that the neurosciences show that the self and ourselves are illusions (Metzinger 2009; Hood 2012). But my brief survey now allows us to appreciate why a novel *positive* position about the self, and hence your nature, actually plausibly follows from the neuroscientific development. Thus, for example, prominent neuroscientists like Panksepp and Biven (2012) or Michael Gazzaniga (2005) are increasingly defending the answer to the Foundational Question that the self, and hence you, are identical to the expansive brain that episodically remembers, hopes, fears, and experiences a body.

Once again, I can quickly reconstruct an argument that drives this contention. We all routinely assume the truth of what I call the "Thinker Thesis":

(Thinker Thesis) You are identical to the individual in your chair that instantiates the properties of remembering, fearing, and thinking.

However, as I have just outlined, and as these neuroscientists are well aware, the neuroscience revolution supports the conclusion that the brain in your chair instantiates the properties of remembering, fearing, and thinking. Plausibly there is also only one individual in your chair that has rich psychology. So, applying the Thinker Thesis, the obvious conclusion is that you are identical to a certain brain. Since brains and animals are not identical, we can thus also conclude that you are *not* identical to an animal. We are consequently led to a new answer to the Foundational Question: you are an expansive brain embodied in a certain animal whose body you control and experience as your own or even yourself.

Evidence for Animalism?

At this point, you are probably writing off my claims as just the kind of implausible speculation that philosophers churn out when we have obvious, and overwhelming, evidence to the contrary. So let me briefly examine the putative evidence supporting the claim that we are identical to animals, for I suggest it now looks questionable. The evidence appears to take two main forms for most of us. First, though rarely discussed by philosophers, there is the evidence of our own lived experience. And, second, as I noted above, there is the putative "fact" that evolution establishes that you are identical to a *Homo sapiens* animal.

The first kind of support is easy to appreciate. Right now you are having an experience of your body, or a "proprioceptive" experience. And you experience certain fingers and arms, for example, as *yours*, as *owned*, or simply just as *you*. But if fingers and arms are you, then you are plausibly a body and hence a *Homo sapiens* animal. We can thus see why our lived experience drives a deeply compelling intuitive judgment in each of us that we are an animal.

In response, however, I need to mark some of the recent findings that the neuroscience revolution has supplied about our proprioceptive experience. First, we now know that the ownership aspect of our experience is a representational feature that can be turned off. Thus there are people with brain lesions who experience their limbs but do not experience them as their own. Like all aspects of our experience, we can thus see that ownership is a constructed aspect of this representation. Second, and more important, we have scientific accounts of the characteristic role of the property of proprioceptively experiencing, which, for example, is produced by certain receptors in the limbs, produces various other rich psychological properties, and produces muscle stimulations, in association with other rich psychological properties. But, once again running the type of argument about the productive roles of rich psychological properties sketched above, we may thus conclude that proprioceptive experience is instantiated in the brain. Applying the Thinker Thesis, we can thus use your bodily experience to show you are identical to a brain and *not* an animal. Recent scientific findings thus provide good reasons to think that our intuitive judgments about what we are based upon our lived experience, however strongly they may grip us, are unreliable and should not be trusted.

What about the evidence from evolution? Basically, in the nineteenth and twentieth centuries the rise of evolutionary frameworks in biology led to something like this little argument being used to dispatch the previously dominant view that we are identical to a kind of individual descended from individuals directly created by God:

(1) You are either not evolved and a kind of individual descended from individuals directly created by God, or you are evolved and identical to an animal.
(2) You are evolved.

Therefore, from (1) and (2):

(3) You are identical to an animal.

Against the background of this argument, we can better appreciate why so many scientists and other writers assume that evolution has established that we are animals.

Unfortunately, the problem is that our scientific understanding has recently radically increased in the ways outlined earlier. So premise (1) is plausibly now either known to be false or not known to be true. For given the arguments outlined above, driven by the advances of the neuroscience revolution, it is now a live option that you are an expansive or thinking brain. But brains and other organs are also evolved. Consequently, we can now only safely assume this alternative premise, with three rather than two options, when we run this type of argument from evolution:

(1*) You are either not evolved and a kind of individual descended from individuals directly created by God, or you are evolved and identical to an animal, or you are evolved and identical to a brain.

Slotting (1*) into the argument, rather than (1), we can still knock out the claim that you are an unevolved kind of individual. But using (1*) rather than (1) we can now only conclude that you are identical *either* to an animal *or* to a brain.

So, I suggest, there are also good reasons to doubt that evolution establishes you are an animal. If one takes rich psychology to be central to our natures, as our continuing endorsement of the Thinker Thesis suggests we do, then one can wonder how evolutionary biology alone could ever have resolved the Foundational Question *prior* to the development of scientific accounts of rich psychology? It is therefore unsurprising that the rise of the new sciences of the mind/brain, with their findings about rich psychology, should have such deep implications about how the Foundational Question should, and should not, be answered.

Concluding Remarks

In finishing, I should note that almost every stage of my argument is contentious. My claims about the foundations of neuroscience, my arguments about what we are, or about experience and evolution, and more, will all be disputed. But my goal in such a short piece has not been to establish such claims.

Instead, on the negative side, I hope I have provided a sense of why there are now *scientific* reasons to doubt that the received wisdom, that we are identical to animals, is the correct answer to the Foundational Question—or at the very least to acknowledge that claiming that we are animals requires a defense that is presently all too rarely supplied. On the positive side, I hope I have highlighted the increasingly persuasive reasons in support of the alternative answer to the Foundational Question that we are identical to the expansive brains illuminated by the neuroscience revolution.

Overall, my final conclusion is therefore that the Foundational Question about our deeper natures is scientific and still open—and that we are rapidly piling up empirical findings suggesting we ought to answer the Foundational Question in a manner alien both to our ordinary ways of thinking of ourselves and to the received scientific wisdom.

References

Azevedo, F., L. Carvalho, L. Grinberg, J. Farefel, E. Ferretti, E. Leite, W. Filho, R. Lent, and S. Herculano-Houzel. 2009. "Equal Numbers of Neuronal and

Non-Neuronal Cells Make the Human Brain an Isometrically Scaled-up Primate Brain." *Journal of Comparative Neurology* 513: 532–41.

Bennett, M. R., and P. M. S. Hacker. 2003. *Philosophical Foundations of Neuroscience*. Oxford: Blackwell.

Churchland, P. M. 1981. "Eliminative Materialism and the Propositional Attitudes." *Journal of Philosophy* 78: 67–90.

Churchland, P. S. 1986. *Neurophilosophy: Toward a Unified Science of the Mind/Brain*. Cambridge, MA: MIT Press.

Fodor, J. 1968. *Psychological Explanation*. New York: Random House.

———. 1975. *The Language of Thought*. New York: Crowell.

Gazzaniga, M. 2005. *The Ethical Brain*. New York: Dana Press.

Hood, B. 2012. *The Self Illusion: How the Social Brain Creates Identity*. New York: Oxford University Press.

Metzinger, T. 2009. *The Ego Tunnel*. New York: Basic Books.

Panksepp, J. 1997. *Affective Neuroscience*. New York: Oxford University Press.

Panksepp, J., and L. Biven. 2012. *The Archaeology of Mind*. New York: W. W. Norton.

Pardo, M., and D. Patterson. 2013. *Minds, Brains, and Law: The Conceptual Foundations of Law and Neuroscience*. New York: Oxford University Press.

Tulving, E. 2002. "Episodic Memory: From Mind to Brain." *Annual Review of Psychology* 53: 1–25.

CHAPTER 4

HUMAN ORIGINS AND THE EMERGENCE OF A DISTINCTIVELY HUMAN IMAGINATION

Theology and the Archaeology of Personhood

J. Wentzel van Huyssteen

For a philosophical theologian deeply committed to interdisciplinary dialogue with the sciences, the privilege of being directly involved with the intriguing issue of human origins for the past few years has been both enriching and an extraordinary challenge. Most important, I have learned that archaeology, anthropology, and especially paleoanthropology present us with a very unusual problem of semantic innovation: in looking back to the distant prehistoric past, how does new meaning come to be, and when this happens, how does interpretation and explanation enable us to reconfigure often long-forgotten meanings about our own origins in this distant past? For the French philosopher Paul Ricoeur, this kind of hermeneutical/epistemological venture always involved a radically inter-disciplinary journey through the lengthy route of multiple detours in di-rect dialogue with the human sciences, the natural sciences, philosophy,

and theology (Ricoeur, in Kearney 2004, 124). In these necessary bound-
ary crossings between various reasoning strategies, we come to learn that
various disciplines transversally connect around shared problems. Therefore, in spite of the fact that the staple of evolutionary anthropology,
archaeology, and paleoanthropology has always been material culture, interpretations of the archaeological record go beyond the material and are
profoundly anchored in the integration of input from a diversity of disciplines (Belfer-Cohen and Hovers 2010, 167). In the art of deciphering
indirect meaning about human origins and human evolution, the distant
past is indeed always mediated through an endless process of multilay-
ered cultural, aesthetic, political, historical, scientific, and philosophical
interpretation. And it is this kind of hermeneutics that is fundamentally
important for any approach to human origins and that will shape our interdisciplinary epistemology of what we can know about the history of
our own origins.

In this essay I want to ask whether *human evolution* as such might
provide us with important bridge theories to theological anthropology
and thus to a positive and constructive way of appropriating Darwinian
thought for Christian theology. From a more philosophical point of view,
I am asking whether Darwin's perspective on human evolution can help
us move forward to more constructive, holistic notions of self and personhood, or to what Chris Fowler (2004) has called "the production of
personhood." I will presuppose in this essay what I have argued elsewhere,
which is that in the history of hominid evolution we find surprising answers to the enduring question of what it means to be a *self*, a human person (see van Huyssteen 2006).[1] In fact, what we now know about key
aspects of hominid evolution affirms what Darwin argued for as crucial
aspects of humanness. To this end I would have wanted, ideally, to consider the problem of human evolution, or the archaeology of personhood,
and its broader impact on theological anthropology, by tracking a number of challenging contemporary proposals for the evolution of crucially
important aspects of human personhood that were all of great significance
for Darwin: the evolution of sexuality; the evolution of cognition, imagination, music, and language; the evolution of morality and the religious
disposition. However, because of time constraints I will presuppose here
that the evolution of these crucial aspects of human personhood ultimately converge on the evolution of the human self.

Rethinking Darwin on Human Evolution

Today then, and not surprisingly, scholars from numerous and highly diverse fields are not only addressing the question of what makes us human but are also seeking multidisciplinary input to inform and enhance their answers to this fundamental issue. These questions do not only pertain to empirical questions about what distinguishes humans from their hominid ancestors; they often also refer to a very different kind of question, namely, which of our specific peculiarities give us humans our distinctive "species specificity" and significance. One popular way of defining human distinctiveness, of course, is to make a clear distinction between anatomical and behavioral differences (van Huyssteen 2006, 203). However, the meaning, markers, and justification of human identity and status have fluctuated throughout Western history. Generally, of course, language has been viewed as a crucial marker (Deacon 1997; Mithen 1996; Mellars 1989, 1991; Noble and Davidson 1996; Tattersall 1998, 2002). In addition, conceptions of defining humanness have lately shifted toward our "prosociality," which we share with primates, as well as our unique propensity for imitation (Cartmill and Brown 2012, 182). Also music (Mithen 2006), sexuality (Sheets-Johnstone 1990), and empathy (de Waal 2006, 2009; Sheets-Johnstone 2008; Kirkpatrick 2005) are in the process of being thoroughly researched and hailed as the foundation of not only language, social norms, and morality but also symbolic and even religious behavior.

The other genuinely panhuman trait is the remarkable human capacity for seeing things from someone else's perspective, generally known as *theory of mind*. Humans are indeed strongly disposed to intuitively understand the motivations of others—to the extent that we often see motivations where they do not exist (Cartmill and Brown 2012, 182). This unique ability does, however, give us adaptively valuable insight into the intentions of our friends, enemies, predators, and prey. And, as is well known, both sadism and compassion are neurologically grounded in this disposition (Cartmill and Brown 2012, 182).

For scholars like Agustín Fuentes (2009) and Richard Potts (1996, 2012), the success of humans as a species can be attributed largely to our tendency toward extreme alteration of the world around us. We not only construct material items, but we engage in the creation and navigation of social and symbolic structures, space, and place in a manner unequaled by other organisms. Most anthropologists would agree that human identity could be seen as interactively constructed by and involved in the construction of a conflux of biological, behavioral, and social contexts (Fuentes 2009, 12). For anthropology to become more familiar with "post-neo-Darwinian" evolutionary perspectives, this kind of mutual enrichment will certainly reveal a significant space for mutual enrichment and overlaps between anthropological and evolutionary approaches. I believe that the same would be startlingly true for interdisciplinary theology.

For this reason, and importantly, some evolutionary anthropologists now find the distinction between "Darwinian" and "neo-Darwinian" unhelpful for many of the current evolutionary theories of interest and argue that we should recognize that there is an expansive body of research and theory that is no longer captured by these headings (Fuentes 2009, 12). Without discounting the important role of natural and sexual selection in biological systems, some anthropologists want to emphasize that scientists are now expanding on Darwin's contributions, and invite us to focus on more recent, emerging trends in evolutionary theory. Notably Eva Jablonka and Marion Lamb's important work, *Evolution in Four Dimensions* (2005), calls for the renewal of evolutionary theory by arguing that we should not focus on just one dimension of evolution, namely, the *genetic*. To this important inheritance system, Jablonka and Lamb now argue for three other inheritance systems that may also have causal roles in evolutionary change: the *epigenetic*, the *behavioral*, and the *symbolic*. Epigenetic inheritance is found in all organisms, behavioral in most, and symbolic only in humans (Jablonka and Lamb 2005, 1–8; Fuentes 2009, 13). This constructivist view moves beyond standard neo-Darwinian approaches and acknowledges that many organisms transmit information via behavior, and thus acquisition of evolutionary relevant behavioral patterns can occur through socially mediated learning. Symbolic inheritance comes with language and the ability to engage in information transfer that can be complex and of high density. What makes the human species so different and so special, and what makes us human, lies in the way we can organize, transfer, and acquire information. It is, therefore, our ability to think and communicate through words and other types of symbols that makes us different (see Jablonka and Lamb 2005, 193–231).

On this view there is much more to evolution than simply the inheritance of genes. Also, the variation on which natural selection acts is not always random in origin or blind to function: new heritable variation can arise as a response to the conditions of life (Harman 2006, 31 f.). Importantly, this interactionist perspective blurs any clear prioritization in inheritance systems and thus requires a clear move away from approaches that are limited to either social or biological focuses. On this view "evolution as construction" is the idea that evolution is never only a matter of developing organisms but of organism-environment systems interacting and changing over time in a dynamic interactive process of *niche construction* as a significant evolutionary force alongside natural selection (Fuentes 2009, 14; see also Ruse 2012, 125).[2] For an understanding of human evolution, this is obviously extremely important: most anthropologists would agree that humans are constructed by and involved in the construction of contexts that are simultaneously physiological, behavioral, historical, social, and symbolic. In this sense human behavioral evolution must be seen primarily as *a system evolving* rather than as a set of independent or moderately connected traits evolving (Fuentes 2009, 15). As such, niche construction is a core factor in human behavioral evolution. The startling conclusion, however, is that we should consider the potential impacts of a diverse array of processes that affect inheritance and evolutionary change and the possibility that natural selection can occur at multiple levels and may not always be the only, or main, driver of change (Fuentes 2009, 16).

In addition, anthropologists, in a move important for theologians, have largely rejected the antiquated dichotomy of nature versus nurture in favor of dynamic understandings of social, biological, and historical complexities. In fact, anthropologists can show that the line dividing the social and the physiological is fairly arbitrary, that no human action or morphological trait exists in a vacuum, and that human history is the conjunctural and emergent product of social, physiological, morphological, symbolic, and historical interactivities (Fuentes 2010, 512). Against this kind of background, then, it should come as no surprise that on the specific matter of human evolution an anthropologist like Jonathan Marks could argue that instead of seeing ourselves as "upgraded" versions of our ancestors, we should accept that we have evolved into biocultural ex-apes. In fact, to imagine that we are "nothing but apes" and to find human nature there actually constitutes a denial of evolution. As Marks (2010, 513) succinctly puts it: we have evolved, get over it! Evolution is indeed the interactive production of difference and novelty, and we are indeed not our ancestors anymore. What we, also we theologians, need is an understanding of evolutionary anthropology that helps us grasp what it means to be a cultural as well as a natural being with remarkable symbolic propensities (Marks 2010, 513).[3]

Crucial to our ability for symbolic behavior is our equally remarkable ability for imagination. From a philosophical and theological perspective, it is exactly at this point where the evolution of the moral sense and of the religious disposition become crucially important. To approach and understand these defining traits, especially the propensity for religious imagination, Fuentes has suggested an important distinction: the quest for understanding the human propensity for religious imagination can be aided and enriched by investigating more fully the core role of the evolutionary transition between *becoming human* and *being human* (Fuentes 2014, 1; Mithen 1996).[4] I believe, however, that in order to begin to understand the emergence of religion, it would also be important to find interdisciplinary points of connection across explanatory frameworks whose focus lie outside the limits of just one specific set of explanations of religion and any one specific religious tradition (Fuentes 2014; van Huyssteen 2006).[5]

In my book *Alone in the World: Human Uniqueness in Science and Theology*, I argued, from an evolutionary point of view, for the *naturalness of religious imagination* (van Huyssteen 2006, 93 ff.). If indeed there were an evolutionary naturalness to religious imagination, or to the propensity for religious belief, then it would be a valid question to ask *how* such an imagination as a system emerged over the course of human evolution. Against the background of a broader, more robust view of the many dimensions of evolution that included extensive, interactive niche construction, we can indeed say that *Homo sapiens sapiens* is a species that had a hand in making itself. From this follows the central theses of the anthropologist Agustín Fuentes's work: Fuentes *first* argues that an evolutionary assessment of a distinctively human way of being in the world includes the capacity and capabilities for the possibility of metaphysical thought as a precursor to religion; *second*, this can be facilitated by recognizing the increasingly central role of niche construction,[6] systemic complexity, semiotics, and an integration of the cognitive, social, and ecological in human communities during the Pleistocene era, that is, roughly two and a half million to twelve thousand years ago (Fuentes 2014).

Following up on my own quest to understand the naturalness of the propensity for religious imagination, Fuentes believes this idea can be aided significantly, as mentioned earlier, by investigating more fully precisely the core role of the evolutionary transition between *becoming human* and *being human*.[7] This transition itself can be understood better by a broad assessment of hominin[8] evolution over the past six million years. And here the focus should be on the terminal portion of that epoch, meaning the final transition from the archaic form of our genus *Homo sapiens* into the current form of *Homo sapiens sapiens*. The focus on this transition, which is a shift to a wholly human way of being in our current sociocognitive niche, will add to our insight into how we, as humans, experience the world in the here and now. Fuentes now suggests that we can connect this emergence of a distinctly human sociocognitive and ecological niche to existing in a meaning-laden world, and to the emergence of an imagination that facilitates the capacity and capabilities for the possibility of metaphysical thought. Moreover, this process is intricately connected to our success as a species (Fuentes 2014, 2).

While many scholars have proposed that the origin of religion and of religious belief is either an adaptation/exaptation or by-product of our cognitive complexity, others suggest that it is more complicated than that (Wildman 2009; Barnard 2012; Donald 2001; Sosis 2009; van Huyssteen 2014). In addition, Fuentes argues that evolutionary answers to the question of the origin of such systems might not lie wholly in the content of religious beliefs or in neurological structures themselves but rather (at least partially) emerge out of the way in which humans successfully negotiated the world during the terminal stages of the Pleistocene (Fuentes 2014, 3). Already the evolutionary epistemologist Franz Wuketits (1990, 118) could argue that metaphysical belief is the result of particular interactions between early humans and their external world and thus results from specific life conditions in prehistoric times. More important, in this evolutionary context one can now envision a distinctive imagination as a core part of the human niche that ultimately enabled the possibility of metaphysical thought. It is ultimately this component of our human niche as our way of being in the world that is the central aspect of our explanation for why *Homo sapiens* has flourished while all other hominids, even members of our own genus, have gone extinct.

On this view, then, looking at human origins and the archaeology of personhood, and thus at the evolution of our lineage across the Pleistocene, it is evident that there is significant increasing complexity in the way we interface with the world (Fuentes 2014, 9). Finally, the emergence of language and a fully developed theory of mind with high levels of intentionality, empathy, moral awareness, symbolic thought, and social unity would be impossible without an extremely cooperative and mutually integrated social system in combination with enhanced cognitive and communicative capacities as our core adaptive niche. Interestingly, on this point Fuentes (2014, 10) wants to incorporate an analysis on *compassion*. I believe this can be pushed even farther back by tracing the deep evolution of empathy and attachment (van Huyssteen 2014; Hrdy 2009, 82 ff.; Kirkpatrick 2005; Sheets-Johnstone 2008). Our genus thus provides a scenario wherein we can envision a distinctively human imagination as a key part of our niche and as a part of the explanation for why our species succeeded and all other hominins went extinct. Fuentes puts it rather forcefully: the imagination and the infusion of meaning into the world by the genus *Homo* in the late Pleistocene is what underlies and preceded our current ability to form a metaphysics, which in turn eventually facilitates religious beliefs.

Fuentes is here in agreement with Terrence Deacon (1997), Merlin Donald (2001), Barbara King (2007), Alan Barnard (2012), and Andrew Robinson (2010) that it is our place as *semiotic species* and the use of symbol as a core infrastructure of our perceptions and of our dealing with the world that act as a major factor, and thus as a hallmark of, human evolution. Humans have an imagination that is part of our perceptual and interactive reality and is a substantive aspect of lived experience. Thus it is realistic to accept that at some point in the past four hundred thousand years language and hypercomplex intentionality acted to lock in the more-than-material as our permanent state of being, and so laid the groundwork for the evolution of morality, the possibility of metaphysics, religious imagination, and the propensity for religious belief as crucial parts of the uniquely human experience. Now, existing in a landscape where the material and social elements have semiotic properties, and where communication and action can potentially be influenced by representations of both past and future behavior, implies the possession of an imagination, and even something like "hope," *the expectation of future outcomes beyond the predictable* (Fuentes 2014, 13). The assertion here is, then, that this interactive process occurs as a component of the human niche as it moves dynamically through the Pleistocene as part of the emerging human tool kit. Imagination, and therefore religion, on this view, is not just an exaptation, a spurious by-product of evolution, but crucial to the process of human evolution and incorporates behavioral processes and a sense of imagination and hope that would, and did, increase the likelihood of innovation and successful responses to evolutionary challenge.

Evolutionary Biology and Theology

This brief overview of human origins and human evolution demonstrates the path and substantive impact of changes in behavior, life histories, and bodies in human ancestors and humans themselves. From this it is clear that patterns that in the Upper Paleolithic would lead to the unambiguous appearance of "art" and "symbol" now also combined with the evolution of empathy and compassion and the deep caring for others. It should therefore not be surprising that a distinctively human imagination is part of the explanation for human evolutionary success and can be seen as one of the structurally significant aspects of the transition from earlier members of the genus *Homo* to ourselves.

A better understanding of cooperation, empathy, compassion, the use of and engagement with materials, symbols, and ritual, and the notion of a semiotic landscape in which humans and our immediate ancestors exist(ed) do indeed move us along in our analysis of what it meant to become a human person. The understanding of all this is indeed a true interdisciplinary process: the insights we gain from the fossil and archaeological record, and behavioral, neurological, and physiological systems, provide a more robust understanding of how humans perceive and experience the world. And it is this process that creates the possibility for an imaginative, potentially metaphysical, and eventually religious experience

of the world (Fuentes 2014, 17). This should lead to a better understanding of the ubiquitous importance of the propensity for religious imagination and the reality of religious experiences for *Homo sapiens sapiens*.
Fuentes is here, correctly, I would say, not arguing for any particular adaptive function of religiosity; rather he argues that in an evolutionary context neither religion nor religiosity could suddenly appear full blown, and
it is therefore valuable to search for the kinds of structures, behaviors, and
cognitive processes that might facilitate the eventual appearance of such
patterns in human beings. If having an imagination is a central part of the
human niche and this imagination is a basal element in the development
of metaphysics, one could see how both adaptive and imaginative, creative perspectives could employ that fact as part of their understanding of
the human.

I believe that my original intuition that there is a naturalness to
human imagination, even to religious imagination, that facilitates engagement with the world in some ways that are truly distinct from those in
other animals—even closely related hominins—thus becomes even more
plausible. In Agustín Fuentes's words: "if this is indeed the case, it provides a small, and hopefully fruitful, addition to the tool kit of inquiry
for both evolutionary scientists and interdisciplinary theologians interested in reconstructing the long, winding path to humanity" (Fuentes
2014, 18).

It should be clear now why, as a Christian theologian interested in
human origins and the controversial issue of human distinctiveness or
"uniqueness," I have been increasingly drawn to the contributions of anthropologists and archaeologists to the challenging problem of what it
means to be human. In my recent work I have been deeply involved in
trying to construct plausible ways for theology to enter into this important interdisciplinary conversation.[9] An interdisciplinary approach, carefully thought through, can help us identify these shared resources in
different modes of knowledge so as to reach beyond the boundaries of our
own traditional disciplines in cross-contextual, cross-disciplinary conversation. It can also enable us to identify possible shared conceptual problems as we negotiate the porous boundaries of our different disciplines.

One such shared interdisciplinary problem is exactly the concern for
"what makes us human," for human distinctiveness and how that may, or

may not, relate to human origins and the evolution of religious awareness. It is, therefore, precisely in the problem of "human distinctiveness" that theology and the sciences may find a shared research trajectory. An interesting part of our self-perception is that it is often the less material aspects of the history of our species that fascinates us most in the evolution of modern humans. We seem to grasp at an intuitive level that capacities like language, self-awareness, imagination, consciousness, moral awareness, symbolic behavior, and mythology are probably the defining elements that really make us human (Lewin 1993, 4). Yet exactly these elements that most suggest humanness are often the least visible in the prehistoric record. For this reason paleoanthropologists correctly have focused on more indirect but equally plausible material pointers to the presence of the symbolic human mind in early human prehistory. Arguably the most spectacular of the earliest evidences of symbolic behavior in humans are the Upper Paleolithic cave paintings in southwest France and in Basque Country that were painted toward the end of the last Ice Age. The haunting beauty of these prehistoric images, and the creative cultural explosion that they represent, should indeed fascinate any theologian interested in human origins. Moreover, I believe that the European Upper Paleolithic era represents a quite spectacular final transition from the archaic form of our genus *Homo sapiens* into the current form *Homo sapiens sapiens*. The focus on this transition, which is a shift to a wholly human way of being in our current sociocognitive niche, will add to our insight into how we, as humans, experience the world in the here and now.

In the interdisciplinary conversation between theology and the sciences the boundaries between our disciplines and reasoning strategies are indeed shifting and porous, and deep theological convictions cannot be easily transferred to philosophy, or to science, to function as "data" in foreign disciplinary systems. In the same manner, transversal reasoning does not imply that scientific data, paradigms, or worldviews can be transported into theology to there set the agenda for theological reasoning. Transversal reasoning does mean that theology and science can share concerns and converge on commonly identified conceptual problems such as the problem of human uniqueness. These mutually critical tasks presuppose, however, the richness of the transversal moment in which theology and evolutionary anthropology may indeed find amazing connections and overlapping intersections on issues of human origins and uniqueness.

Furthermore, I believe that the most responsible Christian theological way to look at human uniqueness requires, first of all, a move away from esoteric, abstract notions of human uniqueness and, second, a return to radically embodied notions of humanness, where our sexuality and embodied moral awareness are tied directly to our embodied self-transcendence as creatures who are predisposed to religious belief. I would further argue that, also from a paleoanthropological point of view, human distinctiveness has emerged as a highly contextualized, embodied notion that is directly tied to the embodied, symbolizing minds of our prehistoric ancestors as physically manifested in the spectacularly painted cave walls and portable art of the Upper Paleolithic. This not only opens up the possibility for converging arguments, from both theology and paleoanthropology, for the presence of imagination and religious awareness in our earliest Cro-Magon ancestors, but also for the plausibility of the larger argument: since the very beginning of the emergence of *Homo sapiens*, the evolution of those characteristics that made humans uniquely different from even their closest sister species, that is, characteristics like consciousness, language, imagination, moral awareness, symbolic minds, and symbolic behavior, has always included religious awareness and religious behavior.

The paleoanthropologist Ian Tattersall (1998, 201) has argued exactly this point: because every human society, at one stage or another, possessed religion of some sort, complete with origin myths that purportedly explain the relationship of humans to the world around them, religion cannot be discounted from any discussion of typically human behaviors. There is indeed a naturalness to religious imagination that challenges any viewpoint that would want to see religion or religious imagination as an arbitrary or esoteric faculty of the human mind. What has emerged from the work of scientists like Steven Mithen, William Noble and Iain Davidson, Merlin Donald, Ian Tattersall and Terrence Deacon, and should be of primary interest to theologians working on evolutionary anthropology, is that human mental life includes biologically unprecedented ways of experiencing and understanding the world, from aesthetic imagination and experiences to spiritual contemplation—exactly the point now being made by Agustín Fuentes about niche construction. Also Deacon (2003, 504) has made the important point that the spectacular Upper Paleolithic imagery and the burial of the dead, though not final guarantees of shamanistic or religious activities, do suggest strongly the existence of sophisticated symbolic reasoning, imagination, and a religious disposition of the human mind. The symbolic nature of *Homo sapiens* also explains why mystical or religious inclinations can even be regarded as an essentially universal attribute of human culture (Deacon 1997, 436), and opens up an interesting space for Jean Clottes and David Lewis-Williams's argument for a shamanistic interpretation of some of the most famous of the Upper-Paleolithic imagery (Lewis-Williams 2002; Clottes and Lewis-Williams 1998). There is in fact no culture that lacks a rich mythical, mystical, and religious tradition. The coevolution of language and brain not only implies, however, that human brains could have been reorganized in response to language and the environment in a dynamic process of niche construction but also alerts us to the fact that the consequences of this unprecedented evolutionary transition from *becoming human* to *being human* must indeed be understood on many levels.

The idea that religious imagination should not be seen as an isolated faculty of human rationality and that mystical or religious inclinations can indeed be regarded as an essentially universal attribute of the human mind has recently also been taken up in interdisciplinary discussion by some theologians (e.g., Shantz 2009). Also, Niels Henrik Gregersen has argued that imagination, and therefore also religious imagination, is not an isolated faculty of human rationality but can be found at the very heart of human rationality. On this view, then, the same "naturalness" of imagination also applies to religious imagination, and religious imagination should not be seen as something extra or esoteric that can be added, or subtracted, from other mental states (Gregersen 2003, 23). More important, though, a theory about the emergence of religious imagination and religious concepts, of course, does not at all answer the philosophical question about the validity of religion, or the even more complex theological question whether, and in what form, religious imagination might refer to some form of reality. As an *interdisciplinary* problem, however, the reasons that may undergird the unreasonable effectiveness of religious belief and thought may transcend the scope of any one discipline when it comes to evaluating the integrity of religious belief. In this specific conversation we can hopefully reach an interdisciplinary agreement that religious imagination and religious concepts should be treated equally with all other sorts of human reflection. Religious imagination should, therefore, be treated as an integral part of human cognition, not separable from our other cognitive endeavors. Clearly early human behavior is not understood if we do not take this religious dimension into account.

I suggest that a theological appropriation of these rich and complex results of science at the very least should inspire the theologian carefully to trace and rethink the complex evolution of the notion of human distinctiveness, or the *imago Dei*, in theology. Theologians are now challenged to rethink what human uniqueness might mean for the human person, a being that has emerged biologically as a center of self-awareness, identity, and moral responsibility. In theology I would, therefore, call for a re-visioning of the notion of the *imago Dei* in ways that would not be disembodied or overly abstract but that instead acknowledges our embodied, evolutionary existence, our close ties to the animal world and *its* uniqueness, and to those hominid ancestors that came before us while at the same time focusing on what our symbolic and cognitively fluid minds might tell us about the emergence of an embodied human distinctiveness, consciousness, and personhood and the propensity for religious awareness and experience.

In my own recent work I have argued that theologians should be intensely aware of how interpretations of the *imago Dei* have indeed varied dramatically throughout the long history of Christianity (van Huyssteen 2006, 111–62). This notion of self or personhood, when reconceived in terms of embodied imagination, symbolic propensities, and cognitive fluidity, will now enable theology to revision its own notion of the *imago Dei* as emerging from nature itself, an idea that does not imply any superiority or a greater moral value over other animals or earlier hominins. In this kind of interdisciplinary conversation theology can actually help to significantly broaden the scope of what is meant by "human distinctiveness." *Homo sapiens* is not only distinguished by its remarkable embodied brain, by a stunning mental cognitive fluidity expressed in imagination, creativity, linguistic abilities, and symbolic propensities. As real-life, embodied persons of flesh and blood we humans are also affected by hostility, arrogance, violence, ruthlessness, and cunning, and therefore are inescapably caught between what we have come to call "good and evil."

This experience of good and evil and theological distinctions between evil, moral failure, sin, tragedy, and redemption lie beyond the empirical scope of the fossil record, and therefore beyond the scope of science (van Huyssteen 2006, 325). It certainly is our evolutionarily developed bodies that are the bearers of human distinctiveness, and it is precisely this embodied existence that confronts us with the realities of vulnerability, sin, tragedy, and affliction. For any scientist drawn to the more holistic niche, of the full dimension of meaning in which *Homo sapiens* has existed since its very beginning, it is precisely theology that may provide a key to a more holistic understanding of the tragic dimensions of human existence and also why religious belief has provided our distant ancestors, and us, with dimensions of hope, redemption, and grace.

Interdisciplinary Conclusions

Thus it finally becomes possible to tentatively ask questions about how these perspectives from evolutionary anthropology, archaeology, and philosophy might inform our contemporary notions of human personhood, how they enlighten our views of human evolution, and also, finally, what their implications might be for theological anthropology, and quite specifically for the iconic notion of the *imago Dei*? As possible answers to these complex questions I propose a few brief abstracts in the form of the following theses.

i) The strong interdisciplinary convergence of theology and the sciences on the question what it means to be human presupposes arguments from both evolutionary anthropology and paleoanthropology, not only for the presence of religious awareness in our earliest prehistoric ancestors, but also for the plausibility of the larger argument: since the very beginning of the emergence of *Homo sapiens*, the evolution of those characteristics that made humans distinctively different from even their closest sister species, that is, characteristics like consciousness, language, imagination, and symbolic minds and behavior, most probably always included some form of religious awareness and religious behavior. Presupposed in this argument, however, is the remarkable degree of adaptability and the versatility of our species. *Homo sapiens* indeed emerged as a result of its ancestral lineage having persisted and changed in the face of dramatic environmental variability and having coped so successfully with interactive niche construction. It is this versatility that also gives new depth to the kind of human symbolic capacities that Jablonka and Lamb (2005) and Fuentes (2009, 2010, 2014) have highlighted in their recent work.

ii) In our thinking about the emergence of religion or of spirituality in prehistory and in our considering the historical human self as *Homo religiosus*, we should not expect to discover some clearly demarcated, separate domain that we could identify as "religion" as such. What this means is that we should avoid making easy and uncomplicated distinctions between natural and supernatural and material and spiritual when trying to understand the long history of the prehistoric self as it hovers between *becoming human* and *being human*. The history and archaeology of the human self demand a more interactive, holistic approach where not just special artistic objects and artifacts, but daily material life itself must have been deeply infused with imagination and spirituality. This suggests that we theologians, along with evolutionary anthropologists and archeologists, can indeed recognize the spiritual or religious in early time periods only through the material legacy of the people of that time. Imagery, sculptures, paintings, and other artifacts, as well as mortuary practices, may not always be exclusively religious but may certainly point to normal living spaces, and practices, as possible symbolic, religious realms.

iii) On this view it should not be far-fetched to link this symbolic material world, so typical of imagistic modes of religion, to the broad spectrum of neurasthetic abilities and religious dispositions of the uniquely symbolizing human mind. For me this sense of a deep material/spiritual entanglement makes it safe to assume that the neurological functioning of the human brain, including important traits like hyperactive agency detection (HADD), empathy, attachment, theory of mind, and altered states of consciousness, like the structure and functioning of other parts of our bodies, is a human universal and that at least some of the material from prehistory clearly suggests an early imagistic, deeply religious culture, of which important experiential and ritual elements would have been carried through to later, more doctrinal modes of religiosity.

iv) I believe that various scholars (cited in van Huyssteen 2006, 2010b, 2014) have made good arguments for the fact that religion in itself is not adaptive. We humans do not possess, as part of our evolved neurological and psychological architecture, intuitive mechanisms designed by natural selection specifically for the purpose of generating religious beliefs or behavior as a solution to particular adaptive problems. However, distinctive neurological traits like empathy, theory of mind, attachment, altered states of consciousness, HADD, and the evolution of the moral sense/intuitive morality should all be seen as part of a much broader niche in which many domain-specific mechanisms have been co-opted in the service of religion and religious belief. Religion thus activates attachment processes but also many other processes like altered states of consciousness and HADD, and it is most probably this combination that is responsible for the widespread success and staying power of religious belief (Wildman 2009).

Theological Conclusions

i) In my book *Alone in the World?* (2006), I argued that the concept of the *imago Dei*, in its many historical and contemporary incarnations and interpretations, has always in some broad theological sense functioned to express the relationship between creator and creatures, God and humans. What we have learned from the history of this canonical tradition is that the idea that humans are created in the image of God never should be argued in abstraction from the concrete historical, cultural, and social ways we find ourselves in today's world. On a postfoundationalist view this calls for a very consciously pluralistic and interdisciplinary approach to theological reflection.

ii) At the heart of the idea of being created in the image of God we find as the deepest intention of the Genesis texts the conviction that the mythical "first humans" should be seen as the significant forerunners of humanity that define the special relationship between God and humans. Being created "in God's image" in a very specific sense highlights the extraordinary importance of human beings as walking representations of God, in no sense superior to other animals, and with an additional call to responsible care and stewardship of the world, also to our sister species in this world. The multilevel meaning of the notion of the *imago Dei* in the ancient biblical texts was, however, transversally integrated into the dynamics of one crucial text: in Genesis 3:22a we read, "Then the Lord said, 'See, the human being has become like one of us, knowing good and evil.'" Clearly, in this important text is embedded the most comprehensive meaning of the biblical notion of the *imago Dei*. Here, in the emergence of an embodied moral awareness, and a holistic, new way of knowing, lies the deepest meaning of the notion of the image of God. Moreover, the theme of the image of God in the texts of the New Testament also reflects a remarkable continuity with the Old Testament texts, where Jesus is now identified as the one who, like the primal human before him, defines the relationship between humanity and God. Furthermore, this notion of the *imago Dei* is as contextual and embodied as that of the "first Adam": what we know of God, we know only through the story of the suffering and resurrection of the embodied person of Jesus, the Jew. Against this background the notion of the *imago Dei* still functions theologically to express a crucial link between God and humans, and should give Christian theologians *intra*disciplinary grounds for redefining notions of evil, sin, and redemption within Christian theology.

iii) An analysis of the many diverse incarnations of the notion of the *imago Dei* in recent theological history reveals the danger of disembodied, abstract interpretations of some of the most important and influential substantialist, functionalist, relationalist, existentialist, and eschatological interpretations of this canonical concept. On the other hand, finding a transversal connection between the best of intentions of these many models for imaging God can take the notion of the *imago Dei* out of the twilight zone of abstraction and reveal it as a powerful symbol and a source of direction for human life. An imaginative, embodied interpretation of the *imago Dei* specifically directs us toward a recognition of the fact that our very human disposition or ability for ultimate religious meaning is deeply embedded in our species' symbolic, imaginative behavior, specifically in religious ritual as that specific embodiment of discourse with God and with one another. This view presupposes the fact that the embodied human person, as "created in the image of God," has biologically emerged in evolutionary history as a center of self-awareness, religious awareness, and moral responsibility.

iv) Feminist theology has crucially influenced any contemporary rethinking of the idea of the *imago Dei* and has unequivocally shown that this doctrine has traditionally also functioned as a source of oppression and discrimination against women (van Huyssteen 2006, 126–32). Any attempt to re-vision the powerful resources of the *imago Dei* should, therefore, specifically uncover the fact that this important theological symbol does indeed give rise to justice and thus exemplifies a root metaphor for understanding the human person. Such an understanding should ground further claims to human rights, because all humans are equally created in the image of God. In his vision for an intercultural theology, George Newlands has argued in similar fashion for the radical ethical dimension of all interdisciplinary work in theology and science. Newlands (2004) develops the same strong ethical dimension implicit in the idea of humans created in the image of God and applies it directly to theological anthropology and interdisciplinary theology. On this view, now enhanced by Christology, Newlands (2006) can argue that a theology (and science, for that matter) that does not build communities in ways that enhance humanity, fails as Christian theology. He is able to go even further and claim that an ethics of care and solidarity implies care for and solidarity with the marginalized at a fundamental, interdisciplinary level. Thus he opens up a creative way to help us recognize that the issue of human personhood and human rights belongs at the heart of any discussion of the *imago Dei*.

v) A postfoundationalist approach to the problem of human uniqueness in theology thus reveals not only a more holistic, embodied way to think about humanness, but also argues for the public voice of theology in our complex contemporary culture. As we find ourselves deeply embedded in the very specific research traditions of our disciplines, a multidisciplinary awareness may now help us also to realize that a particular disciplinary tradition, in this case Christian theology, may also generate questions that cannot be resolved by going back to the resources of that same tradition alone. And it is precisely this kind of interdisciplinary awareness that should lead us across disciplinary boundaries in search of intellectual support from other disciplines.

vi) Rethinking theologically the *imago Dei* as emerging from nature opens up theology to the interdisciplinary impact of the fact that the potential arose in the embodied human mind to undertake science and technology, to create art, and to discover the need and ability for religious belief. It is in this sense that we cannot understand early human behavior, or human personhood itself, if we do not take this fundamental religious dimension into account.

Notes

1. See also van Huyssteen: "Interdisciplinary Perspectives on Human Origins and Religious Awareness"; "What Makes Us Human?"; "When Were We Persons?"; "Post-Foundationalism and Human Uniqueness"; "Coding the Nonvisible"; "The Historical Self."

2. In this synergistic interaction between organisms and their environment *niche construction* emerges as inherently a constructivist process in which biological, ecological, and social/cultural spheres not only interact but also provide a model for human genetic and cultural evolution by incorporating three levels or dimensions: genetic processes, ontogenetic processes, and cultural processes (see Fuentes 2009, 14).

3. Nowadays, of course, scientists assign all extant human beings not just to one species, but to one subspecies, *Homo sapiens sapiens*. All other subspecies have become extinct. The accompanying scientific rhetoric, however, reveals this is no ordinary subspecies. As Tim Ingold puts it, as "doubly sapient," the first attribution of wisdom, the outcome of a process of encephalization, marks it out within the world of living things. But the second, far from marking a further subdivision, is said to register a decisive break from that world. In what many scientists have called the "human revolution," the earliest representatives of the new subspecies were alleged to have achieved a breakthrough without parallel in the history of life, setting them on the path of discovery and self-knowledge otherwise known as culture or civilization (Ingold 2010, 514).

4. A distinctively human imagination is, of course, part of the explanation for evolutionary success. For Fuentes, this means that significant patterns can be found in the evolutionary patterns and processes in the genus *Homo* during the Pleistocene. This is a niche wherein experiences in and perceptions of the world exist in a semiotic context: social relationships, landscapes, and biotic and abiotic elements are embedded in an experiential reality that is infused with a potential for meaning derived from more than the material substance and milieu at hand (Fuentes 2014).

5. See also Sosis 2009, 315–17: "The religious system is an exquisite, complex adaptation that serves to support extensive human cooperation and coordination, and social life as we know it."

6. Regarding the concept *niche:* a niche is the structural and temporal context in which a species exists. As such it includes space, nutrients, and other physi-

cal factors as they are experienced and restructured and altered by the organism and also shaped by the presence of competitors, collaborators, and other agents in a shared environment (Fuentes 2010). The human sociocognitive niche is a cognitive and behavioral configuration that is derived relative to the sociobehavioral contexts of previous hominins. In modern humans it includes cooperation, egalitarianism, theory of mind (mind reading), cultural transmission and innovation, and language. This is a complex and composite niche unique to the human species and is likely a system whose various components emerged during the Pleistocene to reach its current form (Deacon 1997; Fuentes 2014).

7. By "becoming human," Fuentes refers to aspects of human evolution from the appearance of our genus to the emergence of undisputable *Homo sapiens* (150,000–200,000 years ago); by "being human," he refers to evolution in our species since that time (Fuentes 2014).

8. The term *hominin* includes humans and all of those genera and species derived from the lineage that split with the chimpanzee lineage (roughly 7 million to 8 million years ago).

9. Against this background I have argued (2006) for distinct and important differences between reasoning strategies used by theologians and scientists. I have also argued, however, that some important shared rational resources may actually be identified for these very different cognitive domains of our mental lives. Furthermore, it is precisely these shared rational resources that enable interdisciplinary dialogue and are expressed most clearly by the notion of *transversal rationality*. In the dialogue between theology and other disciplines, transversal reasoning promotes different, nonhierarchical but equally legitimate ways of viewing specific topics, problems, traditions, or disciplines, and creates the kind of space where different voices need not always be in contradiction or in danger of assimilating one another but are in fact dynamically interactive with one another. This notion of transversality thus provides a philosophical window onto our wider world of communication through thought and action (Schrag 1992, 148 ff.; Welsch 1996, 764 ff.) and teaches us to construct bridge theories between disciplines while respecting the disciplinary integrity of reasoning strategies as different as theology and the sciences.

References

Barnard, A. 2012. *Genesis of Symbolic Thought*. Cambridge: Cambridge University Press.

Belfer-Cohen, A., and E. Hovers. 2010. "Modernity, Enhanced Working Memory, and the Middle to Upper Paleolithic Record in the Levant." *Current Anthropology* 51 (1): 167–75.

Calcagno, J. M., and A. Fuentes. 2012. "What Makes Us Human? Answers from Evolutionary Anthropology." *Evolutionary Anthropology* 21: 182–94.

Cartmill, M., and K. Brown. 2012. "Being Human Means That 'Being Human' Means Whatever We Say It Means." In "What Makes Us Human? Answers from Evolutionary Anthropology," ed. J. M. Calcagno and A. Fuentes. *Evolutionary Anthropology* 21: 183.

Clark, J. 2008. "Comments." *Current Anthropology* 49 (2): 187–88.

Clottes, J., and D. Lewis-Williams. 1998. *The Shamans of Prehistory: Trance and Magic in the Painted Caves.* New York: Harry N. Abrams.

Deacon, T. 1997. *The Symbolic Species: The Co-Evolution of Language and the Brain.* New York: W. W. Norton.

———. 2003. "Language." In *Encyclopedia of Science and Religion*, vol. 2, ed. J. W. V. van Huyssteen. New York: Macmillan Reference USA.

De Waal, F. 2006. *Primates and Philosophers: How Morality Evolved.* Princeton, NJ: Princeton University Press.

Dewall, F. 2013. *The Bonobo and the Atheist.* New York: W. W. Norton.

Donald, M. 2001. *A Mind So Rare: The Evolution of Human Consciousness.* New York: W. W. Norton.

Fowler, C. 2004. *The Archeology of Personhood: An Anthropological Approach.* London: Routledge.

Fuentes, A. 2009. "A New Synthesis: Resituating Approaches to the Evolution of Human Behavior." *Anthropology Today* 25 (3).

———. 2010. "On Nature and the Human: Introduction" and "More than a Human Nature." *American Anthropologist: Vital Forum* 112 (4): 512–21.

———. 2014. "Human Evolution, Niche Complexity, and the Emergence of a Distinctively Human Imagination." *Time and Mind* 7 (3): 241–57.

Gregersen, N. H. 2003. "The Naturalness of Religious Imagination and the Idea of Revelation." *Ars Disputandi: Online Journal for Philosophy of Religion* 3, www.arsdisputandi.org/.

Harman, O. 2006. "The Evolution of Evolution." *New Republic*, September 4.

Hefner, P. 1998. "Biocultural Evolution and the Created Co-Creator." In *Science and Theology: The New Consonance*, ed. T. Peters, 174–88. Boulder, CO: Westview Press.

Hodder, I. 2011. "An Archeology of the Self: The Prehistory of Personhood." In *In Search of Self: Interdisciplinary Perspectives on Personhood*, ed. J. W. van Huyssteen and E. P. Wiebe, 109–33. Grand Rapids, MI: Eerdmans.

Hrdy, S. 2009. *Mothers and Others: The Evolutionary Origins of Mutual Understanding.* Cambridge, MA: Harvard University Press.

Ingold, T. 2010. "What Is a Human Being?" *American Anthropologist: Vital Forum* 112 (4): 513–14.

Jablonka, E., and M. Lamb. 2005. *Evolution in Four Dimensions: Genetic, Epigenetic, Behavioral, and Symbolic Variation in the History of Life.* Cambridge, MA: MIT Press.

Kearney, R. 2004. *On Paul Ricoeur: The Owl of Minerva.* Aldershot: Ashgate.

King, B. 2007. *Evolving God: A Provocative View on the Origins of Religion.* New York: Doubleday.

Kirkpatrick, L. A. 2005. *Attachment, Evolution, and the Psychology of Religion.* New York: Guilford Press.

Kuijt, I. 2008. "The Regeneration of Life: Neolithic Structures of Symbolic Remembering and Forgetting." *Current Anthropology* 49 (2): 172–88.

Lewin, R. 1993. *The Origins of Modern Humans.* New York: Scientific American Library.

Lewis-Williams, D. 2002. *The Mind in the Cave: Consciousness and the Origins of Art.* New York: Thames and Hudson.

Marks, J. 2010. "Off Human Nature." *American Anthropologist: Vital Forum* 112 (4): 512.

Mellars, P. 1989. "Major Issues in the Emergence of Modern Humans." *Current Anthropology* 30 (3): 349–85.

———. 1991. "Cognitive Changes and the Emergence of Modern Humans in Europe." *Cambridge Archeological Journal* 1 (1): 63–76.

Mithen, S. 1996. *The Prehistory of the Mind: A Search for the Origins of Art, Religion, and Science.* London: Thames and Hudson.

———. 2006. *The Singing Neanderthals: The Origins of Music, Language, Mind, and Body.* Cambridge, MA: Harvard University Press.

Newlands, G. 2004. *The Transformative Imagination.* Aldershot: Ashgate.

———. 2006. *Christ and Human Rights.* Aldershot: Ashgate.

Noble, W., and Davidson, I. 1996. *Human Evolution, Language and Mind: A Psychological and Archeological Inquiry.* Cambridge: Cambridge University Press.

Potts, R. 1996. *Humanity's Descent.* New York: Morrow.

———. 2004. "Sociality and the Concept of Culture in Human Origins." In *The Origins and Nature of Sociality,* ed. R. W. Sussman and A. R. Chapman, 249–69. New York: Walter de Gruyter.

———. 2012. "Environmental and Behavioral Evidence Pertaining to the Evolution of Early *Homo.*" *Current Anthropology* 53 (suppl. 6): S299–S317.

Ricoeur, P. 1992. *Oneself as Another.* Chicago: University of Chicago Press.

Robinson, A. 2010. *God and the World of Signs: Trinity, Evolution, and the Metaphysical Semiotics of C. S. Peirce.* Leiden: Brill.

Ruse, M. 2012. *The Philosophy of Human Evolution.* Cambridge: Cambridge University Press.

Schmidt, K. 2008. *Sie bauten die Erste Tempel.* Munich: Deutscher Taschenbuch Verlag.

Schrag, C. 1992. *The Resources of Rationality: A Response to the Postmodern Challenge.* Bloomington: Indiana University Press.

Shantz, C. 2009. *Paul in Ecstasy: The Neurobiology of the Apostle's Life and Thought.* Cambridge: Cambridge University Press.

Sheets-Johnstone, M. 1990. *The Roots of Thinking.* Philadelphia, PA: Temple University Press.

———. 2008. *The Roots of Morality.* University Park: Pennsylvania State University Press.

Sosis, R. 2009. "The Adaptationist-Byproduct Debate on the Evolution of Religion: Five Misunderstandings of the Adaptationist Program." *Journal of Cognition and Culture* 9: 315–32.

Tattersall, I. 1998. *Becoming Human: Evolution and Human Uniqueness.* New York: Harcourt Brace.

———. 2002. *The Monkey in the Mirror: Essays on the Science of What Makes Us Human.* New York: Harcourt.

Tattersall, I., and K. Mowbray. 2003. "Human Evolution." In *Encyclopedia of Science and Religion*, vol. 1, ed. J. W. V. van Huyssteen, 298–301. New York: Macmillan Reference USA.

———. 2011. "Origins of the Human Sense of Self." In *In Search of Self: Interdisciplinary Perspectives on Personhood*, ed. J. W. van Huyssteen and E. P. Wiebe, 33–49. Grand Rapids, MI: Eerdmans.

van Huyssteen, J. W. 2006. *Alone in the World: Human Uniqueness in Science and Theology.* Grand Rapids, MI: Eerdmans.

———. 2009. "Interdisciplinary Perspectives on Human Origins and Religious Awareness." In *Becoming Human: Innovation in Prehistoric Material and Spiritual Culture*, ed. C. Renfrew and I. Morley, 235–52. Cambridge: Cambridge University Press.

———. 2010a. "Coding the Nonvisible: Epistemic Limitations and Understanding Symbolic Behavior at Çatalhöyük." In *Religion in the Emergence of Civilization: Çatalhöyük as a Case Study*, ed. I. Hodder, 99–121. Cambridge: Cambridge University Press,

———. 2010b. "What Makes Us Human? The Interdisciplinary Challenge to Theological Anthropology and Christology." *Toronto Journal of Theology* 26 (2): 143–60.

———. 2010c. "When Were We Persons? Why Hominid Evolution Holds the Key to Embodied Personhood." *Neue Zeitschrift für Systematische Theologie* 52: 329–49.

———. 2011. "Post-Foundationalism and Human Uniqueness: A Reply to Responses." *Toronto Journal of Theology* 27 (1): 73–86.

————. 2013. "The Historical Self: Memory and Religion at Çatalhöyük." In *Vital Matters: Religion and Change at Çatalhöyük*, ed. I. Hodder, 109–33. Cambridge: Cambridge University Press.

————. 2014. "From Empathy to Embodied Faith: Interdisciplinary Perspectives on the Evolution of Religion." In *Evolution, Religion, and Cognitive Science: Critical and Constructive Essays*, ed. F. Watts and L. Turner, 1035–53. Oxford: Oxford University Press.

Welsch, W. 1996. *Vernunft: Die Zeitgenössische Vernunftkritik und das Konzept der Transversalen Vernunft*. Frankfurt am Main: Suhrkamp Taschenbuch.

Wildman, W. 2009. *Science and Religious Anthropology*. Farnham: Ashgate.

Wuketits, F. M. 1990. *Evolutionary Epistemology and Its Implications for Humankind*. Albany: State University of New York Press.

RESPONSE I

Constructing the Face, Creating the Collective

Neolithic Mediation of Personhood

Ian Kuijt

There is nothing more visual, recognizable, powerful, indeed even individual and personal than the human face. With only a few exceptions, such as identical twins, each human face is unique, serves as a visual signature to others, and exists as a material manifestation of individuality and personhood within our world. Today and through the historical periods, humans have developed words (e.g., portrait) and descriptive phrases (e.g., sour-faced) that highlight the importance of the face as a representation of the self in our world. Our identity and personhood are, in sum, linked to and symbolized by the human face.

In many ways the forager-farmer transition represents the opening of the human mind, the construction of new ways of defining who we are, of organizing social relations, and of shaping identity through material and immaterial means. But there are also important ways in which the development of the world's earliest villages represents the closing of the human mind, a narrowing of options, if you will, and increased tensions between the collective and the individual. In his essay J. Wentzel van Huyssteen argues that the human species is defined by, among other things, our ability to develop symbolic languages, to organize and transfer a high density of information with complex meanings, and to think and communicate through words and other types of symbols. I strongly agree with this view. In the limited space provided here I want to expand this argument in a new direction and explore how our world's earliest agricultural villagers symbolically mediated their identity and personhood within early communities and used clay, stone, and bone to portray their ancestors and materialize personhood through anthropomorphic representation of the human face.

One of the least recognized aspects of the Near Eastern forager-farmer transition, this is the first time when our ancestors systematically and repeatedly created naturalistic representations of the human face (Ibáñez, González-Urquijo, and Braemer 2014; Kuijt 2008; Stordeur and Khawam 2007). There are several important aspects of the emergence of depicting the human face. First, with the development of Near Eastern Neolithic early villages human communities transformed the ways under which personhood was symbolized and materialized through the stunning and remarkable transition from the almost total absence of naturalized representations of the human face to the human face as the dominant focus of expression. Neolithic people for the first time ever created sculptures of human faces often, using a variety of media. Presumably this artistic expression was connected to the development of new forms of social organization with larger-scale village communities and people living in close proximity to each other. Second, there are good reasons to believe that the inhabitants of the world's first villages imagined personhood as being centered on the collective rather than individualized and focused on select people or the individual.

Elsewhere (Kuijt 2008), I have argued that through shared mortuary practices and symbolic representation, southern Levantine Neolithic personhood and identity were linked to the collective at the regional level, not at the local village, household, clan, or individual level. To make this argument let us initially consider anthropological notions of personhood. My starting point is, of course, that personhood is culturally defined and shaped, flexible and self-created. But the social construction of personhood, which I am going to frame as the immaterial and material means by which individuals define their identity and role at different social scales, is also something that has changed with the ongoing development of human cultures.

While there are earlier Upper Paleolithic representations of the human body, it is in the Neolithic period that we see the first major broad interregional focus on the human face. The Neolithic is not just about food; it is about the earliest multiregional, and arguably most important, evolutionary restructuring of human social relationships when people first built villages, became symbiotically linked with the seasonal pulses of wild and newly controlled plants and animals, started to live in high-density buildings, and coalesced into larger groups living in villages. The daily small-scale actions of our ancestors brought about long-term radical shifts in the human trajectory, through which our worlds became increasingly linked to plants and animals and the need for people to live with them. Few events could have had a greater impact on the nature of what it is to be a person, to define human identity as an individual as well as a part of a community.

Constructing the Public: The Neolithic Face and Personhood

Perhaps the most striking materialization of personhood in Neolithic communities is seen in the common construction of detailed faces on plastered skulls. In the Middle Pre-Pottery Neolithic B (MPPNB) period 10,500 years ago, the world's first villagers created and re-created the human face on clay, wood, bone, and stone. Naturalistic representations of the human face, widely viewed by physiologists and anthropologists as a visual foundation for recognizing individuals in our communities, only emerges in human culture around 10,500 years ago in MPPNB villages (Kuijt 2008). As illustrated elsewhere (Goren, Goring-Morris, and Segal 2001; Kuijt and Chesson 2007; Kuijt and Goring-Morris 2002; Verhoeven 2002), in many cases MPPNB people created realistic faces, complete with eyes, nose, and mouth. While there is local variation in the selection of which facial attributes to illustrate, how these were expressed (some eyes were constructed in plaster as closed, while others were defined by insetting seashells), and what technology was used for plastering dif-

Figure 1. Tell Aswad, Syria, plaster skulls CS5, CS3, CS1. From Stordeur and Khawam 2007, fig. 5.1.

ferent parts of the skull (Goren, Goring-Morris, and Segal 2001), in its most basic form skull plastering is an act of reconstructing the body. We see, in short, the use of materials to reconstruct facial features of the living on the physical structure of the dead. In the case of the Jericho, 'Ain Ghazal, and Tell Aswad skulls, for example, MPPNB people reconstructed natural facial features out of clay, with defined eyes, ears, and mouth and perhaps painting of other facial features. At Tell Aswad, Syria, eyes were portrayed as closed and made of clay (fig. 1). These objects were probably used as heirlooms, circulated within and perhaps between households, and likely displayed in public rituals (Kuijt 2008).

In the overwhelming majority of cases in which villagers chose to represent a human face (with eyes, mouth, nose) rather than head, artisans did not identify the gender of individuals. It has been argued (Kuijt and Chesson 2007) that the faces are deliberately homogenized, not historical portraits of individuals and with no differential treatment for females and males. Between the earlier Pre-Pottery Neolithic A (PPNA) period and MPPNB in the southern and central Levant, we witness a remarkable shift from visual focus on the human lower torso and some secondary sexual characteristics to fixated attention to the face, head, and upper torso of the body. While the sample size is very small compared to the later MPPNB, PPNA imagery consists of three general types: rare phalli carved from stone, carved figurines that are full body in scale but without secondary sexual characteristics, or seated figurines, either carved on stone or made of unfired clay. Some of these have breasts but no discernible facial features. Others are focused on the legs and lower torso and, again, do not depict the face and skull. In each of these cases there is a clear identification and specific selection of physical traits.

This pronounced shift may be related to changing views of personhood and identity during this time. Specifically, in the PPNA, the rare cases of figurines recovered by archaeologists appear to focus on the human body—characterized as a stylized representation with the qualities of ambiguity, simplicity, and focus on the lower half. In contrast, figurines and imagery in the MPPNB were focused on naturalistic representations of the face, with detailed treatments of the eyes, nose, ears, and mouth. Just as importantly, the high frequency of these objects in the MPPNB, as well as their large size and use of heirlooms, suggests that plaster skulls and facial representations played an important role in household and community ritual.

The Face and Mediations of the Symbolic Collective

In creating the world's first representations of the human face, artisans in early villages created symbolic abstractions, devoid of gender or personal identity, and with surprisingly little variation. Researchers have debated if the plastered skulls represented historical individuals or a broader anonymous ancestral group (for a range of perspectives, see Amiran 1962; Arensburg and Hershkovitz 1989; Bienert 1991; Ferembach and Lechevallier 1973; Goren, Goring-Morris, and Segal 2001; Hershkovitz et al. 1995; Özbek 2009). Several lines of evidence suggest that plastered skulls were not designed as accurate representations of known people; rather they were stereotyped abstractions. First, as outlined elsewhere (Kuijt 2008), in some cases artisans created faces in clay on defleshed human skulls without the mandible. Second, there are examples where artisans created plaster facial features, such as with Tell Aswad skull No. 14 (Stordeur et al. 2006), which had a nose covering the mouth, a physical arrangement that is unlikely to have occurred in life. Third, artisans portrayed a range of anatomical attributes (e.g., ears) depending on the communities. Thus the presence or absence of clay ears with a specific skull did not indicate that villagers did not have ears so much as the artist decided not to portray them. Fourth, artisans created plastered skulls that were homogenous, with limited variation in facial phenotypes. Examination of the skulls from Tell Aswad, for example, illustrates the remarkable similarity in the appearance of the plastered skulls (see fig. 1). In the case of Tell Aswad, the figures are presented as sleeping, with closed mouths, almost peaceful in appearance. It seems, therefore, that the presence of impossible facial features, as well as the overall but not total homogeneity in facial features, reflects a system of idealized representation, not attempts to represent historical people.

The makers of these skulls were not concerned about accuracy or replicating the facial features of the deceased so much as the representation of certain selected facial features. In this way a new collective face and identity was created using only part of the skeleton and in some cases the skull without the mandible. In the case of the group of seven plastered skulls found in a Jericho phase DI.xlii house, for example, only one had a mandible. In some cases this required the removal of dentition and the compression of facial features on a much smaller physical area (Kuijt 2008). The new plaster mouth, nose, and eyes were shifted, no longer in correct anatomical position in relation to the skull below the plaster. In these cases the chin, which should be located at the lowest section of the mandible, is now modeled from plaster at the bottom of the maxilla. Thus the people who made the decorated skulls embodied life with plaster in a way that stylized the face and simultaneously shifted the entire visual focus upward. What is interesting here is that there is remarkable homogeneity in skull plastering: adults, both male and female, are presented as being the same. Clearly this is not due to a lack of skill among the artisans, for the plastered skulls from sites such as Tell Aswad demonstrate the intentional creation of human faces that were the same. As any artist will attest, it is harder to make faces look the same than it is to make them look different.

Verhoeven (2002) argues that dominant symbolism, such as the representation of the human face, was one of the structuring principles of PPNB rituals and ideology. I would expand this point and argue that the shared MPPNB focus on the face and head was linked to community ideas of memory and embodiment. With the founding of relatively large agricultural villages, mortuary practices and household ritual changed dramatically in several important ways. First, we see the expansion of secondary mortuary practices with the reuse of skulls. This includes the new use of elaborate specialized practices to reconstruct facial attributes on individual human skulls. Second, we see the new appearance of naturalistic plaster skulls, such as at Jericho, and rare stone masks that could have covered the front of a face. Third, in contrast to the PPNA small figurines, in the MPPNB we see the creation and use of large, half-size human statues and busts made of wood, reeds, and plaster. Fourth, we see the appearance of small stone and clay seated figurines. And fifth, we find examples of the construction of small painted heads on the ends of animal bones. While difficult to address through archaeological data, it can be argued that the deliberate focus on the face, in both construction and decoration, as well as the removal of the head of small figurines and the secondary removal of skulls from human skeletons were part of an internally consistent shared system of ritual practices. MPPNB skull plastering should be conceived of as a shared regional system of embodiment with variation in practice based on specific local histories.

With the development of Neolithic villages people crafted their material identity and personhood in such a way that they highlighted collective and shared ancestors. This is an unparalleled moment, for from this point on villagers not only lived in groups, but they also started materializing personhood through collective representations of their symbolic ancestral dead. The emergence of Neolithic villages provides our first clear evidence for an evolutionary shift from small-scale family personhood to multifamily households, neighborhoods, and village-scale conceptualizations. Social relations were constructed on a broad regional scale, and connected to a broader vision of a symbolic collective identity. We see, in short, a significant outward shift from the small scale to the big community, both in terms of new systems of food production and in terms of a reconceptualization of identity and social relationships. If, as argued by van Huyssteen, the human species is defined by our ability to develop symbolic languages with complex meanings, and to think and communicate through words and symbols, then the example of the Near Eastern Neolithic also reflects a new moral framing of the person, one oriented to, at least on some level, membership within the collective, both materially and symbolically.

References

Amiran, R. 1962. "Myths of the Creation of Man and the Jericho Statues." *Bulletin of the American Schools of Oriental Research* 167: 23–25.

Arensburg, B., and I. Hershkovitz. 1989. "Artificial Skull 'Treatment' in the PPNB Period: Nahal Hemar." In *People and Culture in Change: Proceedings of the Second Symposium on Upper Paleolithic, Mesolithic, and Neolithic Populations of Europe and the Mediterranean Basin*, ed. I. Hershkovitz, 115–33. British Archaeological Reports, International, No. 508. Oxford: British Archaeological Reports.

Bienert, H. D. 1991. "Skull Cult in the Prehistoric Near East." *Journal of Prehistoric Religion* 5: 9–23.

Ferembach, D., and M. Lechevallier. 1973. "Découverte de deux crânes surmodelés dans une habitation du VII millénaire à Beisamoun, Israel." *Paléorient* 1: 223–30.

Goren, Y., A. N. Goring-Morris, and I. Segal. 2001. "The Technology of Skull Modeling in the Pre-Pottery Neolithic B (PPNB): Regional Variability, the Relation of Technology and Iconography and Their Archaeological Implications." *Journal of Archaeological Science* 28: 671–90.

Hershkovitz, I., M. Zohar, M. S. Speirs, I. Segal, O. Meirav, U. Sherter, H. Feldman, and A. N. Goring-Morris. 1995. "Remedy for an 8,500-Year-Old Plastered Human Skull from Kfar HaHoresh, Israel." *Journal of Archaeological Science* 22: 779–88.

Ibáñez, J. J., J. E. González-Urquijo, and F. Braemer. 2014. "The Human Face and the Origins of the Neolithic: The Carved Bone Wand from Tell Qarassa North, Syria." *Antiquity* 88: 81–94.

Kuijt, I. 2008. "The Regeneration of Life: Neolithic Structures of Symbolic Remembering and Forgetting." *Current Anthropology* 49 (2): 171–97.

Kuijt, I., and M. Chesson. 2007. "Imagery and Social Relationships: Shifting Identity and Ambiguity in the Neolithic." In *Image and Imagination: A Global Prehistory of Figurative Representation*, ed. C. Renfrew and I. Morley, 211–26. Cambridge: McDonald Institute Monographs.

Kuijt, I., and N. Goring-Morris. 2002. "Foraging, Farming, and Social Complexity in the Pre-Pottery Neolithic of the Southern Levant: A Review and Synthesis." *Journal of World Prehistory* 16 (4): 361–440.

Özbek, M. 2009. Remodeled Human Skulls from Köşk Höyük (Neolithic Age, Anatolia): A New Appraisal in View of Recent Discoveries." *Journal of Archaeological Science* 36: 379–86.

Stordeur, D., B. Jammous, R. Khawam, and E. Morero. 2006. "L'Aire funéraire de Tell Aswad (PPNB)." *Syria* 83: 5–28.

Stordeur, D., and R. Khawam. 2007. "Les crânes surmondelés de Tell Aswad (PPNB, Syrie): Premier regard sur l'ensemble, premières réflexions." *Syria* 84: 5–32.

Verhoeven, M. 2002. "Ritual and Ideology in the Pre-Pottery Neolithic B of the Levant and Southeast Anatolia." *Cambridge Archaeology Journal* 12 (2): 233–58.

RESPONSE II

Imago Dei and the Glabrous Ape

DOUGLAS HEDLEY

Men at some time are masters of their fates:
The fault, dear Brutus, is not in our stars,
But in ourselves, that we are underlings
                    Shakespeare, *Julius Caesar*, act 1, scene 2

C. S. Lewis declares, "Man's conquest of Nature [one might say in this context, human nature] turns out, in the moment of its consummation, to be Nature's conquest of Man." What is a human being? We are rational animals and as such exceptional: we seem a species apart, uniquely one not fixed by instinct. At the same time we cannot "cut nature at the joints" after Darwin. Freud famously spoke of the three great humiliations: Copernicus, Darwin, and himself. Let us forsake Copernicus and Freud. After Darwin it seems clear that the only coherent definition of a member of *Homo sapiens* is as a member of a population group rather than a quasi-Aristotelian essence. It therefore seems perfectly reasonable to speculate about the relation of our purported "nature" to our own evolutionary history. Man bears, as Darwin claims, notwithstanding "all his noble qualities" and "godlike intellect," in his "bodily frame the indelible stamp of his lowly origin" (Darwin [1871] 1998, 643). A recent article in *Nature* announced that "murder comes naturally to chimpanzees."[1] The research was concerned with the question whether the violence had adaptive value or whether it resulted from extraneous factors such as human interference in, or threat to, chimpanzee habitat. Since chimps and bonobos are our closest biological relatives (we are all great apes), can they provide evidence for the emergence of violence among humans? (Notwithstanding circumstances, bonobos, it would seem, do not kill.) Perhaps one can trace this back to the common ancestor of chimps and humans, five million to seven million years ago.

There are, however, two obvious problems. First, does it make sense to speak of murder in relation to creatures that are following their instincts—however brutal those instincts might seem to us? Since a chimp does not possess a language, and certainly not a conceptual language, is not the talk of "murder" an instance of egregious anthropomorphism? Murder is a concept. Consider the definition of this concept in the legal realm and its problems. In most wrangles, it is the relation of the *actus reus* and the *mens rea* that is decisive. It is not the act alone but the act with the intention that makes murder "murder" rather than "manslaughter." Outside common law territories, there is the idea of *dolus eventualis*, literally, "the cunning of the event," to encompass the willful exploitation of an unintended effect of a deed. Sometimes acts of intentional killings are deemed manslaughter because of mitigating circumstances such as provocation. Yet if one thinks of cases of "manslaughter," ranging from instances of crass driving offenses to, say, assisted suicide, these presuppose considerations of responsibility that are startlingly absent in the case of the chimp. The seventeen-year-old driver may genuinely fail to grasp the devastating impact of high speed in an urban area and cause death. Responsibility may be mitigated by youth and inexperience but cannot be entirely exonerated. A caregiver may agonize over the suffering of a relative or patient and may reluctantly (even harboring grave ethical anxieties) assist a suicide. How can such considerations have any force for the chimp?

Second, what does it mean to argue on the basis of analogies with our closest biological relatives? The fallacy is based on imagining chimps, as Jonathan Marks observes, as walking ancestors. There are biological and genetic fallacies here. The biological lineage is not straightforward. Humans are descended from apes unlike chimps in distinct ways. And even if chimps represent the closest analogue to humanity in most respects, in others they do not (Marks 2003). Chimps are intelligent animals, although their brains are much smaller than those of humans. With our large-headed babies, even the birth of humans is a social activity. Whereas chimps kill anyone approaching newborn offspring, humans require immediate help with neonates. The family seems to be the foundation of human social structures. As is often noted, kinship bonds such as aunts, husbands, and grandmothers do not exist among apes. It may be linked to the social structures that humans require for the rearing of their babies, which are born in a comparatively helpless state and in need of years of support and enculturation.[2]

It is a trivial truth that we are animals. But is it more than stating the obvious? Does the claim that humans are animals occlude significant facts about human nature? Yet even if our ape ancestors were sufficiently close to modern *Homo sapiens* to warrant such analogies, it is not clear what the philosophical import of this is. We are descended from apes but also descended from fish. We do not think of birds as therapod dinosaurs even if their roots in the Mesozoic era are fairly clear. If we are happy to go back six million years in order to explain human behavior, why not even further?

The force of *Homo homini lupus est* relies on the striking *metaphor* of man as wolf. Since Darwin claims in *The Descent of Man* that "there is no fundamental difference between man and the higher mammals in their mental faculties" and the differences are of degree, not of kind, should we be remotely surprised by acts of human savagery? (Darwin [1871] 1998, 67). Indeed there has been burgeoning research on the evolutionary history of *Homo sapiens*, especially in relation to the philosophical and empirical questions about "human nature." Is it an exploded anachronism or the crux of serious anthropology? The debate tends to be between the essentialist and the constructivist. We can avoid the Scylla and Charybdis of an unbending genetic determinism of the kind that has become so popular since the Human Genome Project at the end of the 1990s or the relativisms of the social constructivists, relativisms that defy the striking kinship and commonality of human beings. I propose a theological-metaphysical alternative. We are forged of the same stuff as all other creatures and, as such, share the similitude of the created order with the author of all things. Here we can rely on the role of the religious imagination. God is a transcendent spiritual being, and humanity can have communion with him.

Steven Pinker's *The Blank Slate: The Modern Denial of Human Nature* (2002) and Jesse Prinz's *Beyond Human Nature: How Culture and Experience Shape the Human Mind* (2012) reveal the conflict between a universal theory of humanity based on psychology and biology and the opposing claim that cultural diversity defies such universalism or essentialism.[3] Pinker represents a strong form of programmatic naturalism in his evolutionary psychology. The universal and innate "nature" of human nature is constituted by the modular mechanisms, which evolved during the Pleistocene. On this paradigm, the mind is a mechanism forged in the African Pleistocene and history the cultural efferversence of a hard-wired biology. Human beings respond to features in culturally invariant ways—smells, sights, sounds—and these are the result of evolved psychological mechanisms. These mechanisms are discrete in order to solve particular problems (the famous Swiss army model of the mind). Evolutionary biology is an extension of evolutionary neurobiology: the human mind responds to the environment in patterns determined by the modular adaptations that transcend time and culture (O'Hear 1997). Prinz counters such innatism with arguments about our biocultural nature. We have evidence of tools dating back 2.5 million years, and it would seem that our thumbs coevolved with tools. Hence biological dexterity emerged along with technological developments. Another example is menopause. Other apes are fertile until death. It is puzzling that human females should have decades of life while being infertile. The biocultural can be seen in the grim examples of the Romanian orphanages. Children deprived of proper *cultural* input fail to develop properly *biologically*. The orphans of the Ceauşescu era were stunted in their brain development by the cruel destitution of the state orphanages in communist Romania (Nelson, Fox, and Zeanah 2014).

Our ancestors in East Africa one hundred thousand years ago were biologically indistinguishable from us: with chins and foreheads. (The jaws, perhaps, were more powerful.) We have barely changed physiologically in the past hundred thousand years. Since the cranium has not developed meanwhile, it is plausible to assume that the brain has remained constant. Yet for tens of thousands of years nothing very much happened with humanity. Even though anthropologists have been keen to stress the differences between *Homo sapiens* and Neanderthals, the similarities are striking (e.g., brain size and physiognomy). The behavior seems to have been similar, for example, the practice of hunting and the use of hearths. There is some debate about the coloring of bones, but generally there is a marked absence of symbolic activity. The distinctive difference lies in the symbolic rather than the chin or the physique, the invisible rather than the visible, the imagined rather than the perceived. The success of our ancestors, as Jonathan Marks (2012) has argued, was symbolic, not biological.[4]

Programmatic (as opposed to methodological) naturalism assumes continuity with nature, and this is ruptured by the cognitive revolution in 70,000 BC. Up to the emergence of human beings and their rise to preeminence among the animal world, changes were the result of genetic mutation or environmental pressures. From about 70,000 BC, human beings began to impress their collective images and beliefs on the world. Religion, the holy, and the aesthetic are central features of these "imaginings." The paintings of Chauvet are a striking testimony to prehistoric imaginings, in particular, after the shift from the tribes of hunter-gatherers to a pastoral civilization, out of which cities and empires emerged. Amid the technologies of agriculture, irrigation, and construction, nations, gods, and demons emerged. Anthroplogists such as Ian Tattershall note the ubiquity of religion in human society (Tattershall 1998, 200 ff.). Indeed perhaps some inchoate *sensus divinitatis* precedes the emergence of technology.[5] The remarkable site of the prehistoric Temple Gobekli Tepi (10,000 BC) with its 40- to 60-ton T-shaped stones would seem to predate pastoral civilization. These three- to six-meter-high stones have been interpreted as anthropomorphic, and there are various tieromorphic carvings on the stones that would seem to be the greatest monument of the hunter-gatherer period. What, subsequently, of the majesty and sublimity of the human imagination expressed in the Pyramids, the early Buddhas, the Gothic cathedrals, the Alhambra, or the Taj Mahal?

Imago Dei and the Similarity Thesis?

The history of evolution, as is well known, consists of a process of phylogenetic changes in populations over time. The prehistory of humanity makes the definitional question complex. There have been archaic humans (hominids) for two million years. For millions of years hominids were rather unspectacular animals. Our ancestors had large brains but relatively little brawn and were vulnerable to attack from many more powerful and dangerous animals. Perhaps 400,000 years ago hominids started hunting large animals. About 100,000 years ago it would seem that our ancestors moved from being prey to the greatest predator on earth: *Homo sapiens sapiens*. The initial signs of *Homo sapiens* 300,000 to 200,000 years ago in East Africa did not seem initially to herald any great difference. How should we think of the relation between *Homo sapiens sapiens* and extinct hominids: Neanderthal, Denisovan, *Homo erectus*, *Homo floriensis*, *Homo habilis* (Tattershall 1998)? If *Homo sapiens* mated with Neanderthals and Denisovans, how does this affect our view of the distinctively human? *Homo sapiens* developed radically around 70,000 BC and survived while other hominids died out: *Java soloensis* became extinct 50,000 years ago; Denisovans, 40,000; Neanderthaler, 30,000; and Flores, 12,000. *Homo sapiens* is the uniquely surviving hominid. Yet how far back in the African Pleistocene should one go to find the distinctively human? What about the Australopithecus?

Notwithstanding the indistinctness of the biological category of the species, it seems perplexing to deny the idea of humanity as an exclusive and unique "species." Quite apart from the ethical dimension, there does seem to be a transhistorical and cross-cultural humanity. This is not the citadel of fastidious humanists, those elevated principles of urbane bearers of reason and dignity, but equally those rough-hewn and rude transcultural features that anthropologists dwell on: habits of food and sex, incest and cannibalism. In most cultures females have longer and more elaborate hair.[6]

Up to the emergence of human beings and their rise to preeminence among the animal world, changes were due to genetic mutation or environmental pressures. With the emergence of modern human beings this changes. Removed from the normal causal nexus of all other animals, human beings are language users, free agents, and able to contemplate ultimate ends as well as proximate means. We have a traditional language in the West for this: the "image" of God.

Should this be dismissed as archaic mythic language? Or perhaps the relic of a dogmatic theological and philosophical world that is well lost? The language of the "Image and Likeness" is originally a Hebraism. However, this was transformed through the translation of Hebrew into the Hellenic milieu of the New Testament and preserved in the Imperial period into late antiquity. Ireneas distinguishes between image (εἰκών, *eikōn*, Latin *imago*) and likeness (ὁμοίωσις, *homóiōsis*, Latin *similitudo*). Humanity always has the image, but individual humans have lost the likeness. Clement and Origen take it over. Frequently St. Paul's 2 Corinthians 3:18 is employed in discussions of the image: "But we all with unveiled face, beholding and reflecting like a mirror the glory of the Lord, are being transformed into the same image from glory to glory, even as from the Lord Spirit." The occidental idea of the *imago et particeps Dei* was buttressed by the Delphic oracle as interpreted by the philosophers: "Know Thyself" means "know the Divine element": reason (see Plato, *Timaeus* 41cd; Aristotle, *De anima* 408b 29; Cicero, *Tusc. Disput.*, V.13, §38; Augustine, *Enarrations* XLII, 6). The ancient and medieval authors offer substantive accounts. In some metaphysical sense humanity mirrors the divine essence. On this account there is an ontological reason why men and women are in the image of God, for example, as bearers of a rational soul. Thomas Aquinas writes, "ut consideremus de eius imagine, idest de homine, secundum quod et ipse est suorum operum principium, quasi liberum arbitrium habens et potestatem" (*Summa theologica, Prima Secundae, Prooemium*).

Since the Reformation and the Enlightenment such metaphysical accounts have generally been replaced by more functional models. Van Huyssteen expresses a common sentiment when he writes, "Substantive interpretations of the *imago Dei* have subsequently been replaced by so-called functional interpretations, precisely because substantive views were seen as too static, and too expressive of mind/body dualism" (van Huyssteen 2006, 134). For some theologians the model of the "image" is merely a fact about humankind's relation to God (Middleton 2005). Humanity is related to God because of divine fiat or covenant. There may be no essential reason why man is in the image of God except the (possibly inscrutable) divine will or decision. Or the image may just be a function of human dominance over the rest of the created order (McFarland 2005).

Some contemporary theologians opt for a Christocentric notion of human nature. On this view, we understand human nature purely through revelation rather than through the intrinsic qualities of a human being, or what Alan Torrance, for example, calls a "Christian epistemic base" (Torrance 2012). An obvious problem with the theological definition is that it appears positivistic and stipulative. In a culture like modern Europe, where Christianity still exerts enormous influence but very little dogmatic weight, such a position seems somewhat quixotic. It is unsatisfactory because of the "projectivism" challenge: the familiar objection of Xenophon and Feuerbach that human beings are all too apt to project their longings on the universe. Since Xenophon and Feuerbach, Marx and Freud have offered further reason why it may be more economical to view religious beliefs as projections and illusions rather than as deliverances of supernatural truth.

The Darwinian challenge suggests that functionalism is weakened, especially if evolutionary psychology is correct. Why do we think that humans are special? Why do we think that there is a universal human nature? One answer could lie in evolutionary prowess. These glabrous apes have changed from being relatively insignificant creatures to the most powerful animals in the world. Perhaps we are confusing our unique *power* with a purported universal and distinct human nature. The brain is the most obvious contender for generating human uniqueness, but it is hard to understand why the brain evolved. Its size requires a large skull and that makes childbirth in *Homo sapiens* especially precarious. Hence the potential benefits of the large brain were combined with some significant disadvantages. Yet there are considerable negative side effects of the upright posture. Childbirth is difficult and dangerous because of the limitation of the birth canal in obligate bipedals. Back pain is ubiquitous since posture is more difficult if not on all fours and spinal problems constitute a common problem for humans (especially in the industrialized world). One can speculate that the centrality of posture in the yoga tradition is linked to this special difficulty posed by bipedalism. Mircea Eliade (2009, 54) considered the emphasis on posture in yoga as the attempt to overcome finitude and imitate eternal stillness: "a sign of transcending the human condition."

Arnold Gehlen's concept of humanity as "Mängelwesen," or deficit creature, in *Der Mensch: Seine Natur und seine Stellung in der Welt* of 1940 was pointing to this paradox. The lack of powerful teeth or claws to attack foes, the absence of proper body hair to protect from the weather, and even the relative lack of speed for flight from predators made prehistoric humans precarious creatures. Gehlen observed that we should have become extinct! He was arguing for the principle that the fiction of viewing humans as animals serves to emphasize the relevance of the Prometheus dimension of human life. Culture compensates for our weak instincts and feeble bodily powers.

Another distinguished twentieth-century German philosopher, Hans Jonas, presented three distinguishing elements of human culture: tools, images, and graves (Jonas 1996). The tool has existed for 2.5 million years.[7] Our thumbs seem to have coevolved with tools. Other animals can skillfully manipulate the environment: an eagle, for example, can use winds to enhance its speed in flight, but it cannot transform a landscape like a hominid through fire. We can exploit and transform our habitat and environment through tools. The large skull and brain together with upright posture enables the intelligent use of the hands. Yet Jonas thinks that the image is more important than the tool. The human deployment of the image has its source in the "hidden art in the depths of the human soul" to use Kant's language of imagination (1933, A141–B180–81). Through this hidden art and mysterious power of the imagination, human beings are set loose from the constraints of the immediate environment and can represent entities and events through the mind's eye, not least the afterlife. Jonas notes that the most significant distinction is the awareness of death. Man knows death and reflects on its significance and implications. Jonas argued that the possession of tools, images, and awareness of death distinguishes human from other animals. Yet perhaps the sense of the image is primary (Jonas 2001a). Of course, Jonas wanted to claim, quite correctly, that we shape our environment through tools. We are the only animals that know that our lives are finite and can ask questions about whence and whither. As the Cambridge Platonist Benjamin Whichcote (1742, 186) observed, "It ill becomes us to make our intellectual faculties Gibeonites" (cf. Joshua 9:3–37).[8] It is reasonable for the philosophical pragmatist to claim that human needs may ignite much speculation but that the *sense* of the eternal is one of the deepest human needs. Our encounter with goodness, truth, and beauty as images of the eternal and felt as divine presence is that which the "poet of Christian Neoplatonism" refers as the human "prima voglia and desidero supremo" (Boyde 1983, 287, 265).

The Cartesian Theater: An Outdated Intellectualism?

Aristotle defines the human being as a rational animal. Modern philosophers have toyed with the idea of reason as the slave to the passions. Yet Hume generates Kant. The thinker is not an addition to experience but its precondition. This is at the core of the transcendental unity of apperception: the unified consciousness that is more than a collection of images. It is a unified field of perception that requires a thinker. In this sense our perception of objects presupposes a thinking "I" that can generate a unified field of vision. Is human consciousness an illusion like "the setting of the sun" or an exploded error like the Renaissance theory of the humors? Daniel Dennett (1989, 346) famously ridicules this with his characteristic rhetorical bluster as the Cartesian Theater: "the illusion that there is a place in our brains where the show goes on, toward which all perceptual 'input' streams and whence flow all 'conscious intentions' to act and speak. I claim that other species—and human beings when they are newborn—simply aren't beset by the illusion of the Cartesian Theater."

What might we ask, pace Dennett, is philosophically illuminating about the notion of the "Cartesian Theater"? In "The Nobility of Sight," Jonas presents a vindication of sight as "simultaneous unity" rather than "sequential unity" (Jonas 2001b). Within a field of vision the objects are given "at once" rather than determined by a succession of items. Because of this, vision enables detachment. Through vision we encounter a simultaneous image and thus the contrast between Being and Becoming. Consciousness can thus function as an "inner eye." For humans, the mind does not depend upon the environment. One can think of dreams. The relationship between this inner world and the outer perceived environment constitutes the distinctively human. Jonas (2001b, 54) writes, "Seeing requires no perceptible activity on the part of the object or on that of the subject. Neither invades the sphere of the other." Given that "the object is not affected by our looking at it," he observes, "it is present to me without drawing me into its presence."

There is some recent neurophysiological experimentation that reinforces Jonas's claim that vision "outdoes" the other senses.[9] An example from neuroscience can be used as an illustration of the priority of sight. The "rubber hand illusion" reveals the superiority of sight (or visual dominance) over other senses. In the experiment, the real hand is hidden from sight by a board and a rubber hand is placed in front of the subject as if it were his or her own. The tester strokes the rubber hand at the same time the real hand is being stroked. Watching the rubber hand being stroked at the same time as the rubber hand is enough to convince the brain that the "hand" in its visual field is its own. Thus visual information can override propriocentric information from the muscles and tendons. The brain adopts the rubber hand. In the rubber hand illusion vision trumps our proprioceptive information (Botvinick and Cohen 1998).

Let us take Jonas's "image" to mean not just paintings and art, but the freedom to distance ourselves from the immediate environment, to go "off line" to employ a metaphor from computers. There is a fundamental ambivalence here. The nobility of sight according to Jonas is the foundation of the distinctively human capacity to establish a conception of objectivity. Yet as a (properly cantankerous) pupil of Heidegger, Jonas notes that this "nobility of sight" can encourage us to forget the dimension of "being-in-world," our biological being, our primordial relation to the world that obviously precedes any perception or vision of them. According to Jonas, "The complete neutralization of dynamic content in the visual object, the expurgation of all traces of causal activity from its presentation, is one of the major accomplishments of what we call the image-function of sight." Humans are spectators that can contemplate reality in "detachment from the actual presence of the original object" (2001b, 246). This provides the possibility of those momentous distinctions between "Being" and "Becoming," time and eternity, essence and existence at the core of Western philosophy. As Andrew Marvell writes memorably in "The Garden":

> Meanwhile the mind, from pleasure less,
> Withdraws into its happiness;
> The mind, that ocean where each kind
> Does straight its own resemblance find,
> Yet it creates, transcending these,
> Far other worlds, and other seas;
> Annihilating all that's made
> To a green thought in a green shade. (Marvell 1990, 48).

Coleridge (1983, 305) asserts that the "imagination at all events struggles to idealize and to unify." The decisive contribution of the imagination is in our consciousness of the world as a whole. It is the participation and imaginative engagement with the whole that generates the specifically religious aspect. This image of the world as a whole is closely linked to our emotional reactions, a sense of weal or woe. The mind can turn around upon itself. Humans are self-consciously in the world and thus aware of the world as an arena of free agency: religion and metaphysics are unavoidable. The representation through the image is a feature of the human capacity for choice.

Imago Dei and the Human Imagination

Since the scientific revolution, the transformation of the environment through human ingenuity and mastery has increased mightily. *Scientia potestas!* Lord Bacon's pithy adage points to the benefit of knowledge in evolutionary terms. If "knowledge is power," it seems obvious why our ancestors wished to acquire it. Yet why is imagination so pronounced in *Homo sapiens*? What, indeed, are the adaptive payoffs? It has often been observed that the protracted period of play in human infants can furnish and develop sophisticated cognitive skills. Perhaps, for example, the logic of inferences in imaginative literature encourages fledgling intellects to contemplate myriad possibilities. Further, it might foster recognition of the difference between a real cause and mere correlation. The child can thus select between random occurrences and genuine regularities. The recognition of a cause as a mere conjunction of events generates powers of prediction and the ability to think of invisible properties and counterfactuals based on properties.

This theory, however, presents an outlandish and implausible view of children. Should we image three- and four-year-olds as embryonic scientists exploring the causal world or inchoate philosophers trying to work out the logic of beliefs? The child's imagination is populated by many unlikely and improbable events and companions, and its reasoning is heavily shaped by parents and social groups. It has even been suggested that children are more susceptible to errors of judgment because of their need for authority and guidance and their capacity to dwell in the past, present, and future. Thus the experiments of Tetsuro Matsuzawa (2008) that show the superiority of chimps over children in some intellectual tests, especially memory.

The image is a primary and archaic medium of the human imagination. Jonas uses the example of cave paintings:

> Our explorers [hypothetically coming from another planet to ascertain the presence of men on earth] enter a cave, and on its walls they discern lines or other configurations that must have been produced artificially, that have no structural function, and that suggest a likeness to one another of the living forms encountered outside. The cry goes up: 'Here is evidence of man!' Why? The evidence does not require the perfection of the Altamira paintings. The crudest and most childish drawing would be just as conclusive as the frescoes of Michelangelo. Conclusive for what? For the more-than-animal nature of its creator; and for his being potentially a speaking, thinking, inventing, in short "symbolical" being. And since it is not a matter of degree, as is technology, the evidence must reveal what it has to reveal by its formal quality alone. (Jonas 2001a, 158)

Through painting images such as those in the caves of Chauvet or Lascaux our ancestors could visualize memories and represent imagined states, events, and personae. J. Wentzel van Huyssteen in his Gifford Lectures, *Alone in the World?*, interprets the *imago Dei* in terms of the imagination. He presents religion as a decisive part of the lives of our ancestors living in the Upper Paleolithic. Van Huyssteen sees the paintings are the oldest significant products of human imagination and imbued with religious and mythological meaning: "They tell us about who our direct ancestors were, what they thought, and what they could do. They tell us about imagination, about creativity, about consciousness, about the Creator" (2004, 15). In *Alone in the World?* he writes, "The most spectacular evidence of symbolic behavior in humans—and some of the earliest—can be found in the Paleolithic cave art in southwestern France and the Basque Country in northern Spain. . . . As such, the Upper Paleolithic holds an all-important and intriguing key to the naturalness of the evolution of religion, to the creditability of the earliest forms of religious faith, and to what it means for *Homo sapiens* to be spiritually embodied beings" (2006: xvii). Van Huyssteen's proposal has not gone unchallenged. The first problem relates to the interpretation of images *tout court*. How can one fix the nature of any interpretation of images, especially such ancient ones? Second, the shamanistic hypothesis of David Lewis-Williams, on which Van Huyssteen rests some of his case, has been criticized. Van Huyssteen writes:

> More than simple, decorative pictures, these paintings were gateways to the spirit world, panoramas that, in their trance experience, shamans could enter and with which their own projected mental imagery could mingle in three animated dimensions. . . . [T]he rock face was like a veil suspended between this world and the spirit world. . . . [T]he potency filled paint created some sort of bond between the person, the rock veil, and the spirit world that seethed behind it. (2006, 209)

Language is an obvious difference between humans and other animals. The beliefs that we form about the world are conceptually grounded or suffused. What about the imagination? How useful are the images in the Paleolithic cave? The evolutionary psychologist claims that psychological patterns or traits are modular adaptations. Yet the capacity to construct such images like those in Chauvet changes neither the agent nor the environment. It is barely conceivable that the cave images were produced for functional reasons, like a scarecrow, for example. The scarecrow is a natural likeness of a potential predator, hence its efficacy (Jonas 2001b, 58). The Paleolithic man entering the cave will encounter a range of images that constitute a distinct realm of representations with intentional content. It is this intentional content that makes the images so enigmatic for we do not share the beliefs and practices of our hunter-gatherer ancestors. Artistic activity is one of the earliest defining human characteristics. Is it art? This is projecting false categories into prehistory—or is it? Perhaps we, however, can uniquely reflect on our likeness and through our imaginings we can commune with ultimate reality and mirror its source. The famous line of Meister Eckhart is apposite: "The Eye with which I see God is the same Eye with which God sees me" (*Lectura Eckhardi* 1998, 30). Humanity has created images of the sacred since prehistoric times. These sublime images and imaginings of the transcendent become an instrument of revelation: the real presence of the eternal in human history and culture.

ACKNOWLEDGMENTS: I am grateful to the Department of Religious Studies at the University of Virginia for comments and questions, especially Charles Mathewes. I also wish to thank Agustín Fuentes and Aku Visala.

Notes

1. Available at www.nature.com/nature/journal/v513/n7518/nature13727/metrics/news.

2. Ibid.

3. I am grateful to Aku Visala for his guidance with this literature.

4. Marks 2012. The emphasis on the uniquely symbolic capacities of human beings can be found in Mithen 1996.

5. I am not using the term in the specific manner of John Calvin or Alvin Plantinga.

6. One of the most striking objects of the Paleolithic period, the Venus of Willendorf (or woman of Willendorf), has braided hair. Head hair seems to have coevolved with a certain cultural attentiveness.

7. Since the publication of Jonas's book there has been much research devoted to tool use among chimpanzees. This use is, however, clearly very limited compared to that among humans.

8. The Gibeonites are a people condemned to the menial tasks of wood-cutting and carrying water.

9. I am grateful to Carl Gillett for this reference. On the neurological data concerning variety of illusions of bodily "ownership," see C. Lopez, P. Halje, and O. Blanke, "Body Ownership and Embodiment: Vestibular and Multisensory Mechanisms," *Clinical Neurophysiology* 38 (2008): 149–61.

References

Aquinas, St. Thomas. 1948. *Summa theologiae.* 5 vols. Ottawa: Medieval Institute.

Aristotle. 1999. *A Commentary on Aristotle's "De anima."* Trans. R. Pans. New Haven, CT: Yale University Press. 408b 29.

Augustine. *Enarrationes in Psalmos.* Ed. E. Dekker and J. Fraipont. Turnhout: Brepols, 2013.

Botvinick, M., and M. Cohen. 1998. *Nature* 391, no. 756 (February 19).

Boyde, P. 1983. *Dante Philomythes and Philosopher: Man in the Cosmos.* Cambridge: Cambridge University Press.

Cicero. 1927. *Tusculan Disputations.* Cambridge, MA: Harvard University Press.

Coleridge. 1983. *Biographia Literaria.* Ed. W. J. Bate and J. Engell. Princeton, NJ: Princeton University Press.

Darwin, C. [1871] 1998. *The Descent of Man.* Amherst, NY: Prometheus Books.

Dennett, D. 1989. *The Intentional Stance.* Cambridge, MA: MIT Press.

Eliade, M. 2009. *Yoga: Immortality and Freedom.* Trans. W. R. Trask. Mythos: Bollingen Series in World Mythology. Princeton, NJ: Princeton University Press.

Jonas, H. 1996. "Tool, Image and Grave: On What Is Beyond the Animal in Man." In H. Jonas, *Mortality and Morality: A Search for the Good after Auschwitz,* ed. L. Vogel, 75–86. Evanston, IL: Northwestern University Press.

———. 2001a. "Image Making and the Freedom of Man." In *The Phenomenon of Life: Towards a Philosophical Biology,* 157–75. Evanston, IL: Northwestern University Press.

———. 2001b. "The Nobility of Sight: A Study in the Phenomenology of the Senses." In *The Phenomenon of Life: Towards a Philosophical Biology,* 135–56. Evanston, IL: Northwestern University Press.

Kant, I. *Critique of Pure Reason.* Trans. N. Kemp Smith. London: Macmillan, 1933.

Lectura Eckhardi, Predigten Meister Eckharts von Fachgelehrten gelesen und gedeutet. 1998. Ed. Georg Steer and Loris Sturlese. Stuttgart: Kohlhammer.

Marks, J. 2003. *What It Means to Be 98% Chimpanzee: Apes, People, and Their Genes.* Berkeley: University of California Press.

————. 2012. *The Alternative Introduction to Biological Anthropology*. Oxford: Oxford University Press.

Marvell, A. 1990. "The Garden." In *Andrew Marvell*, ed. F. Kermode and K. Walker, 47–49. Oxford: Oxford University Press.

Matsuzawa, T. 2008. *Primate Origins of Human Cognition and Behaviour*. New York: Springer.

McFarland, I. A. 2005. *The Divine Image: Envisioning the Invisible God*. Minneapolis, MN: Fortress.

Middleton, J. R. 2005. *The Liberating Image*. Grand Rapids, MI: Baker.

Mithen, S. 1996. *The Prehistory of the Mind: A Search for the Origins of Art, Religion, and Science*. London: Thames and Hudson.

Nelson, C. A., N. A. Fox, and C. H. Zeanah. 2014. *Romania's Abandoned Children, Brain Development, and the Struggle for Recovery*. Cambridge, MA: Harvard University Press.

O'Hear, A. 1997. *Beyond Evolution, Human Nature and the Limits of Evolutionary Explanation*. Oxford: Oxford University Press.

Tattershall, I. 1998. *Becoming Human Evolution and Human Uniqueness*. Oxford: Oxford University Press.

Torrance, A. 2012. "Is There a Distinctive Human Nature? Approaching the Question from a Christian Epistemic Base." *Zygon* 47: 903–17.

Whichcote, B. 1742. *The Sermons of Benjamin Whichcote*. Ed. W. Wishart. 4 vols. Edinburgh.

CHAPTER 5

WHAT IS HUMAN NATURE FOR?

GRANT RAMSEY

During the battle of Iwo Jima in June 1944, Private First Class Jackylin Harold Lucas and three other U.S. Marines came under attack while making their way along a ravine. Upon seeing two grenades thrown near the soldiers, Lucas dove onto one grenade and pulled the other under his body, saving his companions from serious injury or death. Lucas survived, but his injuries were so grave that his companions left him for dead (Lucas and Drum 2006).

Lucas's act was one of spectacular and nearly suicidal altruism. What does such an act show us about our nature and the human capacity for good and evil? Perhaps it offers a window onto our true nature—a nature ultimately good, though susceptible to corrupting influences. Or perhaps it seems extraordinary precisely because we are greedy and violent by nature but can, in rare instances, rise above those instincts.

Questions about what human nature is and how we can learn about it are difficult to answer. They are difficult not just because humans are complex creatures whose behavior is deeply embedded in the cultural environment that they are a part of, but also because it is not obvious what a concept of human nature is supposed to do or what it is for. The concept of human nature is often used as a *normative* concept, one that can serve as a guide to action, showing us how we ought to behave. Less common-place is an approach that seeks a *descriptive* account of human nature, one that characterizes what humans do and are disposed to do.

I argue in this essay that the normative and descriptive approaches are at odds and that we should not expect a single concept of human nature to play both roles. Furthermore, there are deep problems with normative accounts. They often ignore or contradict the contemporary scientific worldview, and they often merely reflect biases about how we ought to be and what we ought to do. Human nature in this sense becomes politicized and serves in arguments about the moral status of issues like homosexuality, abortion, or biomedical enhancement. Because of the problems inherent in normative notions of human nature, I offer a descriptive alternative. My alternative attempts to align the scientific study of the human with human nature.

Normative Approaches and Their Discontents

If humans are good by nature, then our goal should be to embrace our nature. If humans are by nature bad, knowing what our nature is could provide us with a warning of what we need to strive to avoid, what tendencies we need to overcome. Let's consider what such a target or warning could be and how we could learn about it.

A normative conception of human nature cannot be derived from simply studying humans and making generalizations about them. Doing this merely tells us what humans are like based on observable features. A normative notion of human nature must be about more than manifest behaviors; it must have a special force, one from which we are able to divine not just what we are, but what we are supposed to be like. The question, then, is, What sources can provide us with insight into a nature of this kind?

Normative conceptions of human nature are often religious in origin. For religions with sacred texts, we could perhaps gain direct insight into human nature by reading these texts. Many Christians, for example, take the Holy Bible and its doctrinal interpretations to be a unique source of knowledge about human nature. Consider the *Catechism of the Catholic Church*:

416 By his sin Adam, as the first man, lost the original holiness and justice he had received from God, not only for himself but for all human beings.

417 Adam and Eve transmitted to their descendants human nature wounded by their own first sin and hence deprived of original holiness and justice; this deprivation is called "original sin."

418 As a result of original sin, human nature is weakened in its powers, subject to ignorance, suffering and the domination of death, and inclined to sin (this inclination is called "concupiscence").

419 "We therefore hold, with the Council of Trent, that original sin is transmitted with human nature, "by propagation, not by imitation" and that it is . . . 'proper to each'" (Paul VI, *CPG* §16).[1]

From this it is clear that due to original sin human nature is weak and that we are lusty (inclined to concupiscence), unjust, and apt to be ignorant and to suffer. And these traits are due not to our imitation of others but come directly "by propagation" from our descent from Adam. If this is our nature, then we might infer that we should acknowledge our ignorance, brace ourselves for the suffering that will come, and be wary of our concupiscence. This is, of course, to assume that concupiscence and ignorance are bad things, though these assumptions may be able to be justified from other passages from the Bible.

Although one can read normative interpretations of human nature from some religious passages, since they at times directly state what we are supposed to do and to avoid doing, there are difficulties with this approach to human nature. One difficulty is that religious texts often fail to provide a univocal, coherent picture of what we are or how we ought to be. For example, consider what the Bible says about the fate of the righteous.

Psalm 92:12: "The righteous shall flourish like the palm tree."
Isaiah 57:1: "The righteous perisheth, and no man layeth it to heart."

It is difficult to know from this whether one should strive to be righteous. There are numerous other examples. Should one act altruistically and not hide this behavior from others?

Matthew 5:16: "In the same way, let your light shine before men, that they may see your good deeds and praise your Father in heaven."
Matthew 6:3–4: "But when you give to the needy, do not let your left hand know what your right hand is doing, so that your giving may be in secret. Then your Father, who sees what is done in secret, will reward you."

Is it sometimes OK to lie or to judge others? Again, the Bible gives no single answer; different passages provide divergent answers.

If we interpret these biblical guides to action as providing insight into what kind of being we are (one that ought to be righteous, say), then they can be seen as a window onto our nature. And if this is the case, even if the Bible were the only sacred text, one would have to pick and choose passages to have a clear picture of who we are, how we should act, and what we should strive to become or to avoid. It would be hard for such picking and choosing not to merely reflect prior ideas about human nature instead of informing them. But the Bible is not the only sacred text—far from it. Christianity is not a unified doctrine itself, and it is certainly not the only religion with a creation story and something to say about human nature. Most religions have at their basis a creation story, one that describes human creation and the mark that this creation has on human nature. The creation story endorsed by the Catholic Church is one among many. The picture given by the great Hindu scripture, the Bhagavad Gita, to name just one example, is quite different from that of the Catholic Church. Thus to base human nature on one of these texts presupposes some justification for why the favored text provides the one true picture of human nature.

If the apparent inconsistencies quoted above are true inconsistencies and if such are a general feature of religious texts, then this internal inconsistency, combined with an absence of justification for holding that one and only one is the source of truths, makes it all but impossible to justify one religion or interpretation over another. Conceptions of human nature derived from religion might therefore often be a mere repackaging of one's prior beliefs and inclinations. This may mean that the normative human nature project is doomed. But before one reaches this conclusion, two other normative approaches should be considered, one with a foundation in evolutionary biology, another based on aesthetic preferences.

Evolution by natural selection produces natural functions or purposes in the world. The heart's purpose is pumping blood, despite the fact that it does many other things, such as making noise, burning calories, and taking up space in the chest. The reason that pumping blood is its purpose is that pumping blood is the trait that was selected for over its evolutionary history. Hearts that made noise or burned calories but failed to pump blood were not selected for; organisms with such hearts would be weeded out of the population.[2]

Hearts and other biological adaptations therefore have a kind of normativity; they have something that they ought to be doing. Perhaps, one might think, we could ground a normative conception of human nature on the normativity of natural selection. The benefit of such a conception of human nature is that it would be normative, yet based on objective scientific criteria, not on subjective interpretations of creation myths.

Despite the appeal of the evolutionary approach, it is not in fact a satisfying source of a normative conception of human nature. One reason is that natural purposes are purposes that parts of an organism have in virtue of the fitness contribution these parts provided to their ancestors. Teeth are for slicing and grinding food because of the contribution to biological fitness that teeth playing this role had for ancestors. More generally, the normative character of natural purposes is of this kind: Parts of organisms have purposes in virtue of the ancestral fitness contribution of the parts to the whole. Thus, it does not make sense to ask what the purpose is of whole organisms. This is not entirely true: some organisms are grouped together into superorganisms that can act as biological individuals—ants in a colony or bees in a hive, for example. In these cases, there are distinct casts of insect, and it makes sense to ask what the purpose is of, say, a medium-sized worker in a particular ant species. But in species without specialized casts, it does not make sense to ask what the purpose is of the entire organism. Thus, if the normative character of whole humans—and not their parts—is sought, the naturally selected purposes will have little to offer.

To deny that human nature cannot be derived from natural purposes is not to deny that natural selection was a strong force in shaping human nature. Our nature is what it is to a large degree because of the selection pressures that operated on our ancestors. Our nature is thus evolutionary in this sense. But just because it is evolutionary in this sense does not

mean that we can read a normative human nature off of the naturally selected function of our parts.[3]

If seeking a foundation for natural purposes in religion or evolution is
unlikely to bear fruit, what other possibilities exist for deriving normative
conceptions of human nature? One possibility is to base human nature on
aesthetic preferences. In this case, we have a conception of how we would
like humans to be and how we would like them to act. We label this
human nature, and the normative force of human nature is thus derived
from the difference between how we are and how we want to be. While I
think that it can be a productive exercise to imagine how we would like
humans to be, since this can provide a target for self-improvement, it
seems wrong to label as human nature this ideal human nature. If anything, human nature should be considered that which we wish to mold
into the ideal, not the ideal itself.

Deriving normative notions of human nature from religion, evolution, or our aesthetic preferences is thus problematic in one or more
ways. This is not to say that we can never derive normative conclusions
about our actions from premises about our nature. But such derivations
would not be derivations from human nature alone. Instead, they would
be based on normative premises plus non-normative premises about our
nature.

If the pursuit of normative human nature is unlikely to bear fruit, we
should instead explore the possibility of a descriptive account of human
nature. Such a conception understands human nature to be about what
humans are like, not about what humans ought to be like.

Essences and the Challenges of Descriptive Projects

Normative conceptions of human nature center on figuring out what
it is we ought to be like, whereas descriptive accounts of human nature
aim at characterizing what we are like. We have seen that there are problems with common ways of arriving at a normative account of human
nature, but there are also challenges to producing descriptive accounts—
challenges so difficult that some philosophers have even called for the
abandonment of the concept "human nature." I argue below that essentialist descriptive projects are indeed problematic but will offer a nonessentialist alternative.

The most straightforward descriptive account of human nature considers "human" to refer to the biological species *Homo sapiens* and "nature" to denote the essential properties of the species—properties necessary and sufficient for membership in *Homo sapiens*. The search for human nature, then, is just to find the essence of our species. In order for this account to work, species must have essences.

The problem is that the origin of evolutionary biology in the nineteenth century marked the demise of species essentialism. Darwin, in his *Origin of Species* (1859), made two important advances. One is the proposal that natural selection could be a strong and creative force in evolution, enabling the production of complex traits exquisitely adapted to their uses. The other is the idea that the history of life has a tree structure: there is one giant tree of life, and all the terminal branch tips represent extant individuals. Branches on this tree represent taxa such as species, genera, families, and phyla. Species and genera represent small branches, whereas more basic taxonomic divisions like phyla represent large, central branches.

Darwin spent the first two chapters of the *Origin* on the topic of variation. This was not merely to argue that there is sufficient variation in nature for natural selection to act on. Instead, Darwin wished to make a more fundamental claim about the nature of variation and the status of the taxonomic categories. For Darwin, the tree structure is real, but we have some liberty in the way we draw circles around the branches and provide them with names. This explains why different taxonomists will describe different numbers of species when studying the same specimens. Although it may be that these taxonomists are making errors, it is also possible that they merely have different perspectives on how one should categorize nature's variation and draw species-level circles about the branches. For Darwin, the species category is not different in kind from higher-level genera or lower-level variations. Instead, variations are the fuel for speciation, and speciation can lead to new genera.

The Darwinian insight about the tree structure of the history of life has carried over into contemporary science. The dominant scientific way of reconstructing and analyzing phylogenetic trees is known as cladistics, and the trees they generate are called cladograms. For cladists, taxonomic names (species names or other taxonomic categories) are properly used to denote a branch point and all the descendant branches.

This all becomes important when we ask why a particular individual is a member of a particular species. The answer given to us by contemporary science is that the individual belongs to the species because it resides within a particular branch of the tree of life. It is not because it has certain intrinsic properties. Foxes generally have fur, four legs, a tail, and two eyes, but bearing these traits is not what makes them foxes. A fox is a fox if and only if it is a part of the fox branch of the tree of life. A hairless or one-eyed fox is just as much a fox as a stereotypical fox. The same is true of humans. Although humans typically speak a language and walk on two legs, no individual must exhibit these species-typical traits in order to be a human. And just as humans can be more or less typical, they do not similarly exist on a scale of more or less belonging to *Homo sapiens*.

The philosopher David Hull (1986) used the fact that species belongingness is based on existence in a phylogenetic tree instead of essential properties to argue that humans have no nature. His argument goes like this: The "nature" in human nature denotes an essence. The human species lacks an essence. Therefore, there is no such thing as human nature.[4]

In the face of Hull's argument, there are two responses one could give. One is to concede the soundness of his argument and try to suppress uses of "human nature," arguing that human nature should go the way of antiquated conceptions of chemical elements like the aether and phlogiston. The other response is to challenge Hull's argument, to argue that the first premise is false: The "nature" in human nature does not need to refer to an essence. Instead, there could be a nonessentialist conception of human nature.

There have been a number of attempts to create nonessentialist accounts of human nature.[5] A central challenge is how to deal with diversity and difference. If each one of us were exactly the same as each other, to characterize one of us would be to characterize all. Such a characterization could thus serve as a way of defining human nature. However, each of us is unique and each of us exists in a unique cultural milieu. To characterize one is to characterize all only if one paints with very broad strokes—strokes so broad that the interesting details of our humanness all but disappear.

There have been two main strategies for dealing with this problem. One is to find a way of picking out a set of core traits, each of which belongs to the majority of humans. This is the central tendency strategy: there is a set of traits that humans tend to exhibit. The philosopher Edouard Machery offers an account of this kind. For him, "human nature is the set of properties that humans tend to possess as a result of the evolution of their species" (2008, 323). Under this view, human nature consists of a set of traits. In this case the traits are delimited by two criteria: (1) they must be possessed by the majority of the species, and (2) they must be due to evolution and not merely learned.

This central tendency approach successfully avoids the pitfalls of essentialism and deals with the challenge of diversity by stipulating that not all individuals need to display a trait in order for it to be considered a component of human nature. But does the central tendency concept of human nature do for us what we want a concept of human nature to do? This, of course, gets back to our question of what a concept of human nature is for. If it is to be descriptive, to characterize our species and perhaps say how it differs from other closely related species, it is not clear that the central tendency approach is best. For one, it ignores traits that exist in only one sex. Female menopause is a quite interesting feature of our species. Why should this be excluded from human nature just because it is not exhibited by the majority of the species? Furthermore, the central tendency approach describes human nature in terms of an array of traits, not in terms of how these traits are related to one another. But an important part of our nature is how the traits are causally or temporally related to one another, which traits tend to cluster together, and what the ordered sequence is in which they are typically exhibited. Because of these limitations of the central tendency approach, I have offered (Ramsey 2013) an alternative to this account, which I describe in the following section.

The Life History Trait Cluster Account

Humans live their lives and exhibit traits over their life histories. Some traits like teeth or language are absent from babies but are usually present in adults. Many traits appear in a reliable sequence: adult teeth after baby teeth, baby talk before words, and words before sentences. By "trait," I mean anything exhibited by the organism, including morphological traits (e.g., facial hair), intentional behaviors (e.g., shaving or talking), and behaviors outside of conscious control (e.g., digesting).

For every individual, there is the life they actually lived and there are the countless lives they could have lived. For each of us, the particular way that we encountered the heterogeneity in the environment has had a profound effect on our life history outcomes. Had things unfolded differently in countless ways, I would not be here writing this essay. In fact, for each of us, there are an infinite number of ways that we could have lived our lives. This set of possible lives, despite being infinite in size, is highly constrained and is unique to each of us. Although I could easily have ended up being a biologist instead of a philosopher, I could not have sprouted gills and taken to the water, nor could I have built a chrysalis and metamorphosed into a butterfly. These outcomes are not a part of my possible life histories.

From the raw material of the set of possible life histories, conceptions of individual nature and human nature can be derived. Individual nature can be defined as *the pattern of trait clusters within the individual's set of possible life histories*. Each individual will have a distinct set of traits furnishing their possible life histories, and these traits will be arranged in a unique pattern. This unique pattern of traits is their individual nature: it defines what is impossible, provides probability values for what is possible, and shows which traits are linked to others and how they are linked.

Individual nature serves as a foundation for human nature. If you combine all the possible life histories from all the individual natures, then you will have an even richer set of possible life histories. The trait cluster patterns in this set constitute human nature. In other words, human nature can be defined as *the pattern of trait clusters within the totality of extant human possible life histories*.

Let's now consider how this account of human nature differs from the central tendency accounts, what advantages it has over such accounts, and what challenges it faces. For a central tendency account, human traits are divided into two categories, those that are part of human nature and those that are not. The life history trait cluster (LTC) account, on the other hand, does not have two categories. All traits that exist in the space of life histories are a part of human nature. This might seem like the downfall of the LTC account—if it is so permissive, then surely it cannot provide any insight about human nature.

To see the payoff of the LTC account, consider what sort of human studies are conducted in the social, psychological, and biological sciences. What scientists are interested in, for the most part, is not whether or not humans possess a particular trait, depression, say. Instead, they are interested in discovering the patterns of expression of this trait: what sorts of genetic traits, lifestyles, or traumatic events are associated with depression? The human sciences, then, concern patterns of trait expression— what the traits are, how they are related to one another, and what their causal relations are. If the human sciences are studying human nature, then human nature should concern patterns of trait expression and should not simply be a bucket of traits. The LTC account precisely fills this role and makes human nature the subject of the human sciences.

If the LTC account allows human nature to be the subject of the human sciences, we should ask whether human nature plays the other roles we might want it to play. One role that human nature plays in the popular media is to explain or make sense of the presence of particular human traits. One sees headlines like "Juvenile Males Are by Nature Violent." The usual implied contrast is that such behavior is due to our nature and not culture. One might (rightly) resist this dichotomy and point out that there are not isolated causal influences "nature" and "culture" that can act independently to cause a particular trait. Instead, behavioral traits are produced as a result of a complex interplay of various factors like heritable material, individual learning, and social learning.

These problems aside, we can ask how we can use human nature to explain the way traits are distributed across our species. Within the framework of the central tendency account, the focal question is whether the trait is a part of human nature. The problem is that the above claim is about a subset of our species, juvenile males. And since juvenile males are a minority, nothing they alone do can be a part of human nature (since, at least under Machery's rendering, the trait must be exhibited by the majority). We could instead ask whether "violence" is a part of human nature, but the answer to this does not shed light on the question at hand. With the LTC account, on the other hand, we can investigate whether the traits "juvenile male" and "violent" are associated. Is there a robust association across cultural contexts, say? If so, what explains this? A concept of human nature that rests on trait associations, then, will lead to more productive answers to questions of this kind.

The study of humans is, to a large degree, the study of our traits and how they are associated. No two humans are alike, but they are similar in what traits they display and how they display them. As the philosopher Paul Griffiths (2011, 328) put it, "Humans have a shared nature in the way that vertebrate skeletons have a shared nature. There is structure to their diversity." The goal of the LTC account is to provide a framework within which we can understand and appreciate this structure.

Before concluding, there is one more concept that I would like to offer, that of *uniquely human nature*. Human nature can play the role of characterizing ourselves, but it can also be used to distinguish us from other species. Although human nature, as defined above by the LTC account, is simply an account of the nature of our species, it is uniquely human nature that can set us apart from other species. I will define uniquely human nature as *the subset of human nature that is unique to Homo sapiens*. Although it is human nature for females to lactate after giving birth, since these traits are robustly associated, this is also true of all mammals. Thus, while it is a feature of human nature, it is not a feature of uniquely human nature.

We now have a suite of concepts. There is the set of life histories that form the basis for individual nature. Then there is human nature, which is built out of individual natures. Finally, we have uniquely human nature, built out of human nature minus the nature of other species. These concepts allow human nature to rise out of a hidden inner core to be revealed as what scientists and humanists are studying when they are studying humans.

Conclusion

We have seen that there are deep difficulties with the normative approach to human nature, and that one should be skeptical of those who are not engaging with the sciences in the pursuit of human nature. The account of human nature I offer above attempts to free human nature from the occult realm and suggests that human nature can be seen in the manifest behaviors of fellow humans; to study human nature is no more than to study the patterns of human traits and their causes.

Extraordinary acts like Lucas diving onto grenades show that such acts are within the domain of human nature. But this does not reveal something hidden within each of us, for what is a part of human nature is not always a part of each individual nature. And such acts are a unique outcome of the particulars of the situation. Only by observing repeated instances of like behavior can we gain deeper insight into how pervasive it is within human nature. Finally, concerning the opening question of whether we are good by nature and corrupted by society, or bad by nature and tamed by society, if one takes the LTC view, these alternatives are difficult to make sense of. There is not a core nature prior to experience that we can point to as being good or bad, for culture, experience, and learning are woven into the fabric of life histories in a way that is impossible to unravel. All one can point to are good and bad outcomes and the dispositions that underlie them. We can investigate differences in how individuals are raised and associate this with a diversity of outcomes, but an unraised or uncultured individual does not exist and is therefore neither good nor bad. This view of human nature is liberating. While it acknowledges the infinite variation within our species, it points to constraints in this variation, and makes human nature observable in all acts, lowly and sublime.

ACKNOWLEDGMENTS: Thank you to Agustín Fuentes and Aku Visala for the work and care they have put into this volume. Thank you to Maya Parson and Michael Deem for taking time to carefully read and comment on earlier drafts of this essay. This essay was completed while I was on a National Endowment for the Humanities–supported fellowship at the National Humanities Center. I thank the NEH and NHC for their support. Any views, findings, conclusions, or recommendations expressed in this essay do not necessarily reflect those of the National Endowment for the Humanities.

Notes

1. www.vatican.va/archive/ccc_css/archive/catechism/p1s2c1p7.htm.
2. See Allen, Bekoff, and Lauder 1998.
3. See Ramsey 2012 for a discussion of this point.

4. The claim that species membership is not based on essences should be distinguished from arguments concerning whether or not species represent a natural kind. See, e.g., Ruse 1987; Boyd 1999.

5. See Downes and Machery 2013 for a recent collection of papers on human nature.

References

Allen, C., G. Bekoff, and G. Lauder. 1998. *Nature's Purposes: Analyses of Function and Design in Biology.* Cambridge, MA: MIT Press.

Boyd, R. 1999. "Homeostasis, Species, and Higher Taxa." In *Species: New Interdisciplinary Essays*, ed. R. Wilson, 141–85. Cambridge, MA: MIT Press.

Darwin, C. 1859. *On the Origin of Species by Means of Natural Selection.* London: Murray.

Downes, S., and E. Machery. 2013. *Arguing about Human Nature: Contemporary Debates.* London: Routledge.

Griffiths, P. 2011. "Our Plastic Nature." In *Transformations of Lamarckism: From Subtle Fluids to Molecular Biology*, ed. S. B. Gissis and E. Jablonka, 319–30. Cambridge, MA: MIT Press.

Hull, D. L. 1986. "On Human Nature." *Proceedings of the Biennial Meeting of the Philosophy of Science Association* 2: 3–13.

Lucas, J. H., and D. K. Drum. 2006. *Indestructible: The Unforgettable Story of a Marine Hero at the Battle of Iwo Jima.* Cambridge, MA: Da Capo Press.

Machery, E. 2008. "A Plea for Human Nature." *Philosophical Psychology* 21 (3): 321–29.

Ramsey, G. 2012. "How Human Nature Can Inform Human Enhancement: A Commentary on Lewens's *Human Nature: The Very Idea.*" *Philosophy and Technology* 25: 479–83.

———. 2013. "Human Nature in a Post-Essentialist World." *Philosophy of Science* 80 (5): 983–93.

Ruse, M. 1987. "Biological Species: Natural Kinds, Individuals, or What?" *British Journal for the Philosophy of Science* 38: 225–42.

RESPONSE I

The Difficulties of Forsaking Normativity

NEIL ARNER

I applaud Grant Ramsey's affirmation that human nature is both a valid subject of scientific investigation and a relevant consideration in the process of making sound ethical judgments. His own conception of human nature as the pattern of clustered antecedent-consequent traits across human life spans provides a highly creative response to some significant objections to this twofold affirmation.[1] My response begins by reviewing Ramsey's description of the scientific tractability and ethical relevance of his concept of human nature. I then identify a tension in Ramsey's account pertaining to the matter of normativity. I conclude by proposing that Ramsey alleviate this internal tension by demonstrating a greater openness to the potentially normative valence of the concept of human nature.

In conformity with the counsel of Agustín Fuentes and Aku Visala in the introduction to this volume, I wish to make explicit from the outset my own metaphysical assumptions. I am a Christian theologian broadly committed to the affirmations of the fourth-century Nicene Creed. Although I briefly mention some theological premises related to the God described in that creed, I do not in any sense offer warrants for adopting them. I therefore do not presume to have shown that my theological concept of human nature in any sense trumps Ramsey's scientific concept of the same object. In this context, I only wish to show that my explicitly normative concept of human nature does not have the same problem of coherence that I discern in Ramsey's non-normative concept. My response to Ramsey's position is therefore an *immanent* critique that grants his presuppositions but seeks to disclose an internal inconsistency in it. In no respect must one share my theological convictions in order to agree with my analysis of Ramsey's position.

Empiricism and Metaphysics

Chief among the "core desiderata" for a concept of human nature, according to Ramsey, is the requirement that the concept be "empirically accessible" and thus in "accord" or "alignment" with the biological sciences broadly construed (PEW 986; IHE 481). His own account satisfies this requirement because it contains "merely descriptions of patterns within the collective set of human life histories" (PEW 988). Ramsey also states that his concept of human nature is "based on scientific laws, regularities, or generalizations" (IHE 479). Ramsey's *metaphorical* description of a theoretical construct that is accessible to/aligned with/grounded on empirical facts leaves some ambiguity about exactly how such alignment or grounding is to be judged. Nevertheless, let us take Ramsey at his word and recognize that he wants an account of human nature that is, in some sense, scientifically credible.

Ramsey is quick to note that his proposed criterion rules out several metaphysical accounts of human nature. First, it renders "questionable" or "scientifically bankrupt" ontologies of human nature based on Platonic forms, Aristotelian teleologies, or "occult essences" (PEW 983, 986; IHE 479). "There is not some external 'human nature,' like a fixed target in Plato's heaven, that humans can strive for. Instead, human nature simply tracks the morphology, behavior, and so on, of humans" (PEW 992). Although the initial criterion of empirical groundedness is somewhat vague, let us again grant to Ramsey that a metaphysically naturalist rendering of that criterion does in fact rule out the metaphysical views he identifies.

Concurrent with Ramsey's rejection of a nonempirical human essence is his judgment that a scientifically adequate concept of human nature is *necessarily* incompatible with having a normative property that "tells us something about what humans should be or should strive for" (PEW 986–87). The concept of human nature developed by Ramsey is therefore not "in any sense 'good'"; it is "strictly speaking descriptive and not normative" (PEW 992). His scientific rendering of human nature "does not entail moral principles" unless integrated with normative premises from some other source (IHE 480). Ramsey therefore forces us to choose between having a scientifically credible account of human nature and having a normatively laden account.

My concern about Ramsey's proposal is captured by a question raised by Jonathan Marks at the conference for which these essays were originally prepared: How does one come to know so much about the metaphysical limits of human nature? For example, do we *really know* that there is no Platonic form of human nature? Ramsey does not make the softer claim that believing in such is inadequately justified; he instead claims that such a form categorically does not exist.

Furthermore, if such a metaphysical truth is in fact knowable, by *what means* can we come to apprehend it? Fuentes and Visala likewise wonder, in the introduction to this volume, "who is to say that there cannot be an account of human nature that is not scientific." A Platonist like Douglas Hedley, another contributor to this volume, would surely object to Ramsey's suggestion that such a view is evidently falsified by scientific considerations.

The issue with which Ramsey is wrestling, "the problem of the criterion," concerns how one can verify one's procedure for generating knowledge without already possessing paradigm cases of knowledge in advance. Ramsey resolves this problem by devising an epistemic method that purports to be metaphysically neutral; he then uses that method to decipher what is real. Yet philosopher Roderick Chisholm claims this "methodist" approach begs the fundamental question and "leaves us completely in the dark so far as concerns what reasons [one] may have for adopting this than some other" (Chisholm 1982, 67).

I can illustrate the error of Methodism with a metaphor. A woman who goes trawling for fish by dragging a net with one-inch-square holes is likely to discover at the end of the day that she caught no fish shorter than one inch in all three of its dimensions. Yet if this person were prone to generalizations and thereafter declared that the sea does not contain any fish smaller than one square inch, then we would surely object to his hasty conclusion. We might reply that this method of scrutinizing the contents of the sea *cannot possibly* catch such objects; suitably adjudicating the existence or nonexistence of these small objects requires some method that has not predetermined the outcome in advance.

In a similar manner, Ramsey drags the net of empiricism through the human population, finds that filter to have captured mere behavioral patterns, and then declares that there is no such thing as a normatively laden final cause, moral ideal, Platonic form, or essentialist human nature. But this outcome has been predetermined by his method of investigation. A net designed only to catch empirical objects will never trap an immaterial object. It thus seems that the valid conclusion of Ramsey's inquiry is not that human nature per se is non-normative. Instead, what Ramsey has shown is that if you presuppose metaphysical naturalism, what you find when examining human practices is mere behavior rather than normative ideals. But note that all of the eliminative work of this conclusion is being accomplished by the *presupposition* rather than by any *evidence* or *argument*. No amount of infographics containing clusters of life trajectories can, by themselves, establish this distinctively metaphysical outlook.

Nature and Normativity

For the sake of contrast with Ramsey's account, let me disclose my own metaphysical assumptions. I think that immaterial objects can be real. In particular, I hold the most important among such objects is the creator, sustainer, and redeemer of all material reality. My theological convictions therefore incline me to affirm that human nature is normative for three reasons. First, God has ordered all things—humans included—toward the ultimate good of God's own self. Therefore, human nature has a purposive direction that is divinely imprinted. Those who want their lives to flourish must orient themselves to this normative final end (Matt. 5:8). Second, God has provided a model of this normative ideal in the first-century life of the divine-human named Jesus. (Celia Deane-Drummond elaborates this understanding of God's "incarnation" in her essay in this volume.) God's willingness to take on human nature indicates the potential excellence of this nature. In its ideal form, human nature can reflect divine glory—this, I take it, is part of what it means for humans to bear God's "image" (Gen. 1:26–27; Ps. 7:4–5; Rom. 8:17–18). Third, humans are under an explicit law—a *nomos*, or norm—that has been legislated by God. The summary of that law is that humans ought to love both God and their fellow humans (Deut. 6:5; Lev. 19:18; Matt. 22:37–40). God will ultimately hold all people accountable for their obedience or disobedience to this demand. Therefore, with respect to a final end, ideal exemplar, and authoritative decree, God establishes the normative cast of human nature. While this nature cannot be exhaustively discerned by strictly empirical investigation, scientific investigation is at least capable of apprehending the broad patterns of human flourishing among which are included the end, ideal, and laws given by God.

As an aside, addressing a moral law reminds me of the one consideration that I wish to add to the collective effort, represented by this volume, to characterize human nature. Just as J. Wentzel van Huyssteen indicates in his remarks in this volume, I propose that humans are distinctively responsible agents: people who are subject to normative constraints and ideals. I think that humans can distinctively be asked to give an account of their actions. Humans imagine themselves to be under moral laws that pertain to all species members as such. Humans are thus distinctively aware of being under a personal obligation—a point that Marks also makes in this volume when he contrasts humans and other primates.

Now let us return to my point of dispute with Ramsey. How should he and I resolve our substantive metaphysical difference over the normative character of human nature? We surely cannot resolve the matter by choosing an *epistemic method* that is neutral between us, for any epistemic outlook presupposes some metaphysical commitment. I think the matter should be settled by remaining open to the possible normativity of human nature. This stance coheres well with the generous plurality modeled by Fuentes and Visala when they state in this volume's introduction, "We hold that what it means to be human cannot be reduced to one single perspective or definition." Furthermore, I think that two considerations of internal consistency should persuade Ramsey to adopt this view.

First, Ramsey employs normative considerations in elaborating his own concept of human nature. Ramsey distinguishes his position from that of Tim Lewens chiefly by rejecting the latter's claim that an empirical concept of human nature "can have no bearing on issues with ethical dimensions like that of human enhancement" (IHE 479). Instead, affirms Ramsey, his own concept can "inform human enhancement" in the sense of "link[ing] existing moral principles with data, enabling us to make informed decisions about human enhancement" (483). This informing consists in identifying which physical modifications of humans consistently lead to "deleterious," "risky," or "beneficial" consequents (482). Ramsey explains the moral relevance of this data about antecedent-consequent relations in the following quotation, which is preceded by his declaration that his account of human nature is "strictly speaking, descriptive and not normative."

> This does not mean that there are no moral implications of human nature. . . . If the study of human nature is the study of patterns of trait associations, then studying human nature may provide insight into human goodness and evil—if a particular nefarious consequent is robustly associated with a particular set of antecedents, then this lends support for the elimination or reduction of one or more of the antecedents. (PEW 992)

In this passage, Ramsey tacitly commits himself to two normative principles, neither of which is justifiable by strictly empirical considerations. In the domain of *metaethics*, he is implicitly committed to a consequentialist account of morality according to which the moral appropriateness of an action depends in some respect on the consequences of that act. In the domain of *metaphysics*, he is implicitly committed to some account of human well-being or proper functioning that provides the standard according to which one can consistently discriminate between the beneficial and nefarious consequences for humans. The very notion of "beneficial" presupposes some teleology: the meaning of a concept requires the context of being beneficial *for* some end. It is curious that this tacit commitment brings Ramsey near to the Aristotelian teleology that he explicitly rejects (IHE 479). Philip Kitcher—whom Ramsey cites appreciatively in his own work (IHE 479)—says in response to another, strictly scientific account of human nature, "The notion of ethical objectivity, ostentatiously expelled through the front door, sneaks back through the rear" (Kitcher 1985, 421).

Ramsey's unacknowledged reliance on a normative account of human nature is not so spooky a move as his comments about exorcizing "occult essences" might suggest (PEW 986). Instead, he is merely using the commonplace notion that we can identify some broad measures of what it is for a life to go well in a cumulative sense.[2] I think that Ramsey is correct to relate human nature to this distinctive form of flourishing; I simply invite him to make that commitment explicit rather than leave it merely tacit.

I think that there is yet another reason that Ramsey's own commitments should preclude him from categorically denying the normative quality of human nature. Ramsey employs agreement with human intuition as one criterion for an adequate concept of human nature (PEW 986). Now what are the most widespread intuitions concerning whether human nature is normative? I am not aware of a comprehensive survey that has provided adequate empirical evidence to resolve this question. Yet I wager that at least a substantial minority—if not a majority—of people take themselves to be under some form of moral law that provides normative direction for their conduct. This sense of being accountable is what Immanuel Kant identifies when he states that humans are awestruck by the "moral law within" (Kant [1788] 1997, 269 [5:161 in *Gesammelte Schriften*]). Among living humans, more than half are theists who take themselves to be obligated by some moral law imposed by God. For all such people, there is a normative implication to human nature. Ramsey's respect for human intuitions should therefore incline him to endorse rather than reject a normative account of human nature.

Concluding Remarks

I do not presume to have proven the opposite of Ramsey's position. Demonstrating that human nature is in fact normative requires more space for argument than I have at my disposal. Yet I do hope to have shown that we should not hastily deny the normative quality of human nature. For one thing, it is quite difficult to avoid tacitly assuming that human nature has some normative character. I suspect that most of us do so, and we might even find that we can admit as much without intellectual embarrassment. I also suggest that one of the most attractive candidates for a normative property of human nature is the quality of being accountable to the moral demands placed upon us by persons with the proper standing to do so. Moreover, I think that both theists and nontheists can credibly endorse this property.[3]

For another thing, many people who plausibly present themselves as otherwise intelligent agents hold that human nature is in some respect normative. Claiming that "we" simply know better today is an example of selection bias, whereby one only counts as worthy interlocutors those who already agree with one's normative eliminativism. There is a real dispute to be had here on metaphysical and metaethical grounds; the matter cannot be resolved simply by assuming metaphysical naturalism to be the only fitting method of discerning what truly exists.[4]

Notes

1. In order to render Ramsey's views as accurately as possible, I will reference the following two of his articles: "How Human Nature Can Inform Human Enhancement: A Commentary on Tim Lewens's *Human Nature: The Very Idea*," *Philosophy and Technology* 25, no. 4 (2012): 479–83, http://dx.doi.org/10.1007 /s13347-012-0087-2 (hereafter cited parenthetically in the text as IHE); and "Human Nature in a Post-Essentialist World," *Philosophy of Science* 80, no. 5 (2013): 983–93, www.jstor.org/stable/10.1086/673902 (hereafter PEW).

2. For a defense of this position, developed in conversation with contemporary philosophy of biology, see Porter 2005, 53–140.

3. For two fine elaborations of this kind of "social theory of obligation," see the respectively theistic and nontheistic accounts provided in Adams 1999 and Darwall 2006, 3–118.

4. For a similar argument concerning the importance of excavating and then disputing the metaphysical premises that undergird purportedly "strictly scientific" accounts of morality, see the 2012 Gifford Lectures by Sarah Coakley, which are currently available in video form (www.youtube.com/watch?v=M6xi YZec1wE&list=PL4DB23FD6CCABD62F). See also the account of "critical

scientific realism" advocated by Fuentes and Visala in the introduction to this volume.

References

Adams, R. M. 1999. *Finite and Infinite Goods: A Framework for Ethics.* New York: Oxford University Press.

Chisholm, R. 1982. "The Problem of the Criterion." In *The Foundations of Knowing*, 61–75. Minneapolis: University of Minnesota Press.

Darwall, S. L. 2006. *The Second-Person Standpoint: Morality, Respect, and Accountability.* Cambridge, MA: Harvard University Press.

Kant, I. [1788] 1997. *Critique of Practical Reason.* In *Practical Philosophy*, ed. and trans. J. G. Mary. New York: Cambridge University Press.

Kitcher, P. 1985. *Vaulting Ambition: Sociobiology and the Quest for Human Nature.* Cambridge, MA: MIT Press.

Porter, J. 2005. *Nature as Reason: A Thomistic Theory of the Natural Law.* Grand Rapids, MI: Eerdmans.

RESPONSE II

Some Remarks on Human Nature and Naturalism

Aku Visala

First, I would like to thank Grant Ramsey for his highly interesting and original view on what human nature might mean. In what follows, I briefly comment on some aspects of Ramsey's view that I find persuasive and useful. As will soon become clear, I am sympathetic to Ramsey's view, because it avoids the pitfalls of essentialist and nativist views of human nature. I then go on to make some critical points about human nature and the assumption of naturalism. I conclude my remarks by reflecting on the theological problematic surrounding naturalism about human nature.

Anti-Essentialism and Nurturism

The main reason I find Ramsey's proposal rather persuasive is because it elegantly circumnavigates two widespread eliminativist arguments. In the introduction, we pointed out that most writers in this volume want to find meaningful and justified uses for the notion of "human nature" instead of eliminating the idea entirely. Several biologists and philosophers

240

have recently voiced their concerns about the usefulness of human nature as a scientific concept. Of these concerns, Ramsey mentions David Hull's much-discussed anti-essentialist argument. However, Hull's is not the only one: many critics of evolutionary psychology have argued that since there are no robustly innate cognitive mechanisms, there is no human nature. In other words, there is no human nature, because there is no biologically or genetically given psychology. We could dub the former the "anti-essentialist argument" from Darwinism and the latter the "nurturist argument." Let us begin with the Darwinian argument.

As Ramsey points out, the enemy of anti-essentialist arguments is the everyday, essentialist view of human nature. In addition to Hull, who Ramsey discusses, there are others who have pushed this objection against human nature essentialism. One is the philosopher John Dupré, who puts the main claim like this:

> What has become increasingly clear to post-Darwinian biologists is that there can be no necessary and sufficient condition for being an organism of a certain species, and the characteristic properties of members of a species are, first, almost always typical rather than universal in the species and, second, to be explained in various different ways rather than by appeal to any simple or homogeneous underlying property. (2002, 155)

What Dupré is saying here is that the categorical understanding of species membership receives no support from evolutionary biology. On the Darwinian view, even if there are in fact shared characteristics in a given species, they might not be and need not be uniform in the population. Most members of the current *Homo sapiens* population have the capacity for walking upright or thinking about mathematics, but these features are not what make the individuals in the population *Homo sapiens*. Instead, they are *Homo sapiens* by virtue of being members of a specific historical population. Dupré's view also highlights the fact that population thinking demolishes essences as explanatory factors: whatever similarity there is across a population of individuals, the similarity might have various causes, and those causes need not invoke any shared, essential factors inherent to the organisms themselves. If Dupré is correct, there is very little hope for a robust unique or universal human nature in biology.

Closely linked to the Darwinian argument is the nurturist argument. In his recent book *Beyond Human Nature: How Culture and Experience Shape the Human Mind* (2012), the philosopher Jesse Prinz argues against Steven Pinker and others. For Prinz, the question of human nature has to do with the relationship of cultural and biological influence on human behavior, that is, whether there is an innate component to human behavior. His main argument against there being such a thing as human nature is that most human capacities and behaviors are not innate in the sense of having a specialized psychological mechanism as a base. Prinz himself defends what he calls a rather thoroughgoing nurturism (as opposed to naturism or nativism):

> The mind comes furnished with few, if any, innate ideas, and the innate rules of thought can be used for a wide range of different cognitive capacities. Most of our specific capacities are learned, and the cognitive differences between humans and our close animals stem largely from small improvements in the general-purpose mechanisms that we share with them. Character traits, vocational dispositions and aptitude may be influences by our genes, but they are also heavily influenced by experience. (2012, 11)

He then goes on to examine various aspects of human cognition, including language, moral emotions, and reasoning, while arguing that no robustly innate "modules" or "specialized mechanisms" need to be assumed in order to explain our performance.

Notice that the anti-essentialist and nurturist arguments locate "human nature" in a different place. On Hull's view, human nature, if it exists, has to do with human uniqueness or distinctiveness. If individual members of the *Homo sapiens* population are not characterized by a set of distinctively *Homo sapiens* traits, there is no human nature. On the nurturist view, human nature is understood somewhat differently. For Prinz, the issue is whether we can meaningfully distinguish traits and cognitive mechanisms that have genetic causes from those that are products of culture and the environment. He then provides reasons for why this is very difficult or impossible.

The major benefit of Ramsey's life history trait cycle account of human nature is that it avoids both the anti-essentialist argument and the nurturist argument. First, it does not assume the existence of some set of essentially *Homo sapiens* properties. Instead, human nature consists of patterns of traits across all possible life histories. Second, unlike Edouard Machery's account that Ramsey criticizes, Ramsey's life history account does not commit its adherent to a strong distinction between innate and sociocultural traits. Although Ramsey offers a way to incorporate some aspects of innateness into his account, no assumptions of how the causal factors are distributed among genes, environment, learning, and experience are needed in principle.

Being Human and the Assumption of Naturalism

Ramsey is admirably clear about his naturalistic starting point. His first desideratum for any account of human nature is that it be "empirically accessible subject of behavioral, biological, economic, and social sciences." This, he maintains, rules out a significant aspect of our folk notion of human nature, namely, the desideratum that accounts of human nature tell us what our goals and values should be. In other words, Ramsey seeks a notion of human nature that is free of any normative burden and is purely descriptive. It also rules out aspects of human nature that are not scientifically tractable or accessible. Accounts invoking metaphysical essences or transcendent goals fail to fulfill this desideratum.

Although I do not share Ramsey's naturalism, I will not attempt to argue against it here. Also, I do not want to imply that it is somehow illegitimate or irrational. What I do want to say is that while the assumption of this kind of naturalism is indeed widespread in some circles of analytic philosophers and scientists, it is nevertheless far from being universally accepted. Naturalism is not the only game in town as far as many philosophers, anthropologists, and theologians are concerned.

Large-scale ontological and epistemological assumptions make a rather sizable difference as to how we understand and approach human nature. This is true of naturalism as it is about various other large-scale worldviews. Let me take just a few examples.

First, commitment to naturalism will strongly constrain how we identify the referent "human" in human nature. As Carl Gillett points out, the Foundational Question is not often even formulated at all. Instead, it is simply assumed (as Ramsey seems to do) that we are identical to *Homo sapiens* organisms. This assumption, however, is far from being universally accepted or self-evident. Instead of anchoring the "human" to a biological category, *Homo sapiens*, the "human" has traditionally been taken to refer to persons, thinking things with certain social, moral, and intellectual standings. Of course, the concepts "person," "personality," "agency," and their cognates are contested among philosophers, theologians, and scientists. Often persons are defined as having certain capacities for agency, intellect, and relatedness that, in turn, ground the person's moral value. For the most part, these capacities have to do with self-consciousness, self-direction, reason, and language use. On this view, we are *Homo sapiens* only accidentally, not essentially. Contemporary variations of this view include Gillett's brain view. Gillett argues that because we are mental, thinking things and brains (instead of *Homo sapiens* bodies) exemplify thinking, feeling, wanting, and other mental properties, it seems warranted to assume that we are identical to brains.

Second, naturalistic assumptions tend to reduce or implicitly change the questions we ask about human nature. If we assume from the outset that human nature must be scientifically tractable, we will end up focusing on the kinds of questions that scientific methods are the most suitable to study. If the only thing you hold in your hand is a naturalist hammer, everything concerning humans will start to look like a nail pretty quickly.

As I already mentioned, Ramsey is admirably clear about his assumptions. This, however, is not true of everyone. Let me take just one example of what I mean by this tendency to implicitly change the question. Consider the essay "What Does It Mean to Be Human? An Evolutionist's View" by David Sloan Wilson (2012). Here Wilson's task is to answer the question of what it means to be human. He thinks that the question should be answered in the same way as the question, "What does it mean to be species S?," and goes on to argue that biologists have at their disposal various tools to answer such questions. For nonhuman species, answering this question involves giving an account of the evolutionary history, environment, selection pressures, and adaptations that resulted in the current population. In other words, what it means to be species S is a purely biological question.

So, although the forces at work in human evolution are somewhat distinct from the causes of the evolution of nonhuman animals, what it means to be human is still understood as a question about the various aspects of the evolution of *Homo sapiens*. Wilson's answer follows predictable lines:

> The answer to the question "What does it mean to be human?" is that co-operation is the signature adaptation of our species. One manifestation of cooperation is a symbolic inheritance system that makes adaptation an extremely fast process, at least compared to genetic evolution. (2012, 20)

In other words, the question of the meaning of being human is, for Wilson, a question about the possible distinct or signature features of human evolution. Since human evolutionary history includes distinct features, like symbolic inheritance systems (cultures) and the flexibility of human behavior, these are what being human is about.

Finally, let me mention why this issue of naturalism is highly relevant for theological approaches to human nature. Traditionally, theologians have adopted a rather thick view of human nature along the lines of the everyday, essentialist view. I think it is fair to say that traditional Christian views have tended to affirm the existence of some capacities and powers that are present in some way in all and only all human beings; that is, the traditional Christian quest for human nature is the quest for a universal and unique human nature. Theological anthropology is somewhat less interested in questions of innateness, though claims to innateness are by no means foreign in theological discourse. The assumption of innate essence, however, seems to be there.

But what is this essence and is it scientifically tractable? More "naturalistically" oriented theologians, like van Huyssteen, think that science and theological anthropology converge (at least to some extent) on human nature. Van Huyssteen takes what he himself calls a bottom-up approach to human nature that starts from the results of human sciences and works itself toward a more general theological anthropology. This approach is not without its critics. Many theologians are cautious of naturalism and strict bottom-up approaches, because they seem to compromise the distinctiveness of theological anthropology. Theology, they argue, should reject naturalistic approaches to human nature, because naturalistic, or bottom-up, approaches are either irrelevant for theology or simply false. Although I disagree with this line of thought, let me examine it a bit further.

One extreme example of this kind of theological anthropology is the Scottish theologian Alan Torrance. His main point is that the theologian has a radically different evidence base for his claims about humans than the naturalist. The most central anthropological resource in the Christian evidence base is, or should be, the doctrine of the incarnation: the person and the life of Jesus Christ are, at the same time, the disclosure of God's nature and love for humanity, and the disclosure of what human nature ultimately is. This, Torrance argues, makes the Christian view incompatible with any naturalist view based on biology or psychology. He writes:

> An epistemic base characterized by Christian theism does not allow us to define 'human nature' with reference to a general (psychological, biological, or physiological) analysis of the human that brackets out our relationship to God and his creative purposes. . . . [T]o conceive of the human being outside of its relation to God is not a neutral option. It is to allow the focus of our analysis to be humanity in a dysfunctional state—a state that distorts its creative telos. (2012, 909)

Let us unpack this a little. First, according to Torrance, the Christian view affirms that human beings are more than products of evolution and resists the attempt to flesh out the essential nature of humans in biological terms. If basic Christian affirmations about creation and Christology are true, then essential human nature cannot be identified by biology or psychology. This is because human beings are not essentially as they happen to be now. More specifically, human beings in their current state are sinful, and dysfunctional; in contrast, human being as God intends—as God is creating and redeeming and sanctifying them to be in the eschaton—is, as Christ is, perfect.

On this view, theological accounts of human nature should not rely on the naturalistic sciences. There is an essential human nature, but we are not there yet: we see glimpses of that nature in Christ, and in the life of his Body, the Church, but it will ultimately be revealed only in the eschaton. This teleology is, according to Torrance, invisible to the sciences: in the case of human nature, theologians have access to special knowledge that is outside the purview of the sciences.

I have now offered three examples of how one's metaphysical and epistemological commitments have a strong influence on how human nature is approached. Large-scale metaphysical and epistemological views have been and are still being debated among philosophers, scientists, theologians, and others, and there is no universal agreement in sight. We should work toward mutual understanding but at the same time acknowledge that perpetual disagreement about such matters might leave us disagreeing about human nature as well.

References

Dupré, J. 2002. *Humans and Other Animals.* Oxford: Oxford University Press.

Prinz, J. 2012. *Beyond Human Nature: How Culture and Experience Shape the Human Mind.* New York: W. W. Norton.

Torrance, A. 2012. "Is There a Distinctive Human Nature? Approaching the Problem from a Christian Epistemic Base." *Zygon* 47 (4): 903–17.

Wilson, D. S. 2012. "What Does It Mean to Be Human? An Evolutionist's View." *Minding Nature* 5 (2): 17–23.

PUTTING EVOLUTIONARY THEORY TO WORK IN INVESTIGATING HUMAN NATURE(S)

Agustín Fuentes

One of the main roadblocks to getting a full suite of disciplines to effectively engage on the topic of human nature(s) is the failure to sincerely read across areas and reasoning strategies. A substantial percentage of the researchers thinking about the human and issues of human nature(s) do not engage with the range of relevant literature from disciplines different from their own. A lack of exposure to what is actually being said within and across these disciplines remains a major obstacle to effective discourse on the subject, and this volume is a small attempt to ameliorate that problem. Interaction between a diversity of disciplines with a stake in human nature(s) can facilitate a move beyond current stalemates toward ongoing, and transdisciplinary, discourse and identify potential theoretical and intellectual concordances.

One area where many of the current impasses have their root is in the conceptualization of what it means to take an "evolutionary" approach in talking and thinking about human nature(s). Often when evolution or evolutionary theory is invoked there is an assumption that it is the traditional neo-Darwinian tool kit and perspectives that are being utilized. This need not be the case, and increasingly it is not. The extended evolutionary synthesis (EES) (e.g., Fuentes 2009; Laland et al. 2014, 2015; Pigliucci and Muller 2010) is emerging as a central contemporary approach in evolutionary biology and theory, and it provides an expansion on the well-established but limited tool kit of the traditional neo-Darwinian perspective. The EES provides a more robust and more dynamic theoretical context for attempts at evolutionarily engaged interdisciplinary, or transdisciplinary, investigations of human nature(s).[1]

In this brief essay I identify what contemporary evolutionary theory via the extended evolutionary synthesis is and how it can combine with select elements emerging from the essays and commentaries in this volume. These connections can provide illustrations of how, and where, we can move the conversation on human nature(s) forward. I end with a few thoughts on transdisciplinarity and why it might also be a particularly well-suited approach for such integrations.

Contemporary Evolutionary Theory

Evolution is always a synergy of multiple processes, and natural selection, one of the processes by which biological variants achieve differential representation in subsequent generations, is not the sole architect of function. In our current understandings of evolutionary biology we cannot ignore the other processes (in addition to selection) as salient in any description of evolutionary histories and patterns.

Basic contemporary knowledge of how biological evolution works can be summarized as follows:

- Mutation introduces genetic variation, which in interaction with epigenetic and developmental processes produces biological variation in organisms, which may be passed from generation to generation.
- Natural selection shapes variation in response to specific constraints and pressures in the environment (*sensu lato*), and gene flow and genetic drift structure the distribution and patterns of the genetic components of that variation.

- Dynamic organism-environment interaction can result in niche construction (the process by which organisms simultaneously shape and are shaped by their ecologies), which can change or shape the patterns, foci, and intensity of natural selection and create ecological inheritance.
- Phenotypic plasticity, developmental plasticity/reactivity, and acquisition of nongenetically induced features (via neo-Lamarckian processes) all can play substantive roles in the patterns and production of variation.
- Multiple pathways of inheritance (genetic, epigenetic, behavioral, and symbolic) can affect evolutionary processes. Evolutionary processes can no longer be reduced to a focus on just genetic material. As the core research group promoting the EES states, "We hold that organisms are constructed in development, not simply 'programmed' to develop by genes. Living things do not evolve to fit into pre-existing environments, but co-construct and coevolve with their environments, in the process changing the structure of ecosystems" (Laland et al. 2014, 162).

Jablonka and Lamb (2005) demonstrate that information, the variation that is the fuel for evolutionary change, is transferred from one generation to the next by many interacting inheritance systems (genetic, epigenetic, behavioral, and symbolic). Epigenetic inheritance, the inheritance of molecular or structural elements outside of the DNA, is found in all organisms. This gives rise to phenotypic variations that do not stem from variations in DNA but are transmitted to subsequent generations of cells or organisms (Jablonka and Raz 2009). Behavioral inheritance is the transmission, across generations, of behavioral patterns or particulars and is found in most organisms; and symbolic inheritance, the cross-generational acquisition of symbolic concepts and ideologies, is found only in humans and can have pronounced effects on behavioral patterns. Variation is also constructed, in the sense that, whatever their origin, which variants are inherited and what final forms they assume depend on various filtering and editing processes that occur before and during transmission.

More than a half century of debate about the targets and levels of natural selection has produced a robust recognition that selection likely acts on multiple levels in social organisms (e.g., genomic, individual, group) (Wilson and Wilson 2007; Laland and Brown 2011). So social behavioral patterns that have outcomes at the level of social groups, and perceptions influencing those patterns, can be as significant in evolutionary processes as are those traits that are tied directly to individuals' potential reproductive success (the key indicator in natural selection) (Sober and Wilson 1998; Nowak and Highfield 2011). In regard to humans, Andersson, Törnberg, and Törnberg (2014) note that "Darwinian forces are seen as necessary but not sufficient for explaining observed evolutionary patterns," and it is argued that the extended evolutionary synthesis is better suited than traditional neo-Darwinian approaches to encompass aspects of social-cultural systems as central in evolutionary processes (Read 2012).

Asking about the Human

Behavioral and social inheritances play particularly salient roles in evolutionary patterns for many primate species (Campbell et al. 2011; Strum 2012), especially members of the genus *Homo*, the human lineage (Henrich 2011; Kendal 2012; Andersson, Törnberg, and Törnberg 2014). Given this, it is important to incorporate multiple evolutionarily relevant processes of inheritance (not just genetic but also epigenetic, behavioral, and symbolic/cultural) into evolutionary models and as part of any philosophical deployment of evolutionary theory (Jablonka and Lamb 2005). Recognition of multiple modes of inheritance and a feedback dynamic between organisms and their ecologies is a central tool in integrating behavioral, biological, and ecological factors in modeling human evolution (O'Brien and Laland 2012; Flynn et al. 2013).

Laland, Kendal, and Brown (2007) illustrate that constructing and inheriting socioecological contexts via human material culture (tools, clothes, towns, etc.), and niche construction in general, can occur through cultural means. O'Brien and Laland (2012) demonstrate this by reviewing the evolution of dairying by Neolithic groups in Europe and Africa and the rise of the "sickle-cell allele" among certain agricultural groups in West Africa. I review work on early hominin toolmaking as another example of niche constructive processes in human evolution (Fuentes 2015). In these examples shifting behavioral actions, cultural perceptions, and ecological conditions are shown to have mutually interacted to produce genetic and physiological changes that themselves result in further modification to behavior, physiology, and ecologies. Cultural patterns and behavioral actions and perceptions can have an impact on genetic and other biological processes and the functioning natural selection, which in turn can affect developmental outcomes (e.g., Richerson and Boyd 2005; Henrich 2011), which can then feed back into the cultural patterns and behavioral actions, continuing the dynamic interface.

Our current understanding of evolutionary processes suggests that behavioral/cognitive plasticity combined with increasingly essential modes of social cooperation and coordination enabled humans to develop our modern capacity for extensive shared intentionality, meta-coordination, and language (Sterelny 2012; Tomasello 2014). Indeed, this capacity and proclivity for cultural complexity is increasingly invoked as a key to evolutionary explanations of human behavior (Henrich 2011; Kendal et al. 2012; Richerson and Boyd 2005), and thus is most certainly a basal starting point for evolutionarily engaged inquiries about human nature(s).

In light of contemporary understandings of evolutionary processes and the advances in human evolutionary studies over the past decade, we can assert a few basic assumptions about evolutionary approaches to understanding humans: (a) human evolution is a system not best modeled via a focus on individual traits or genic reductionism; (b) feedback, rather than linear, models are central in modeling human dynamics, so niche construction is important; (c) ecological and social inheritance are significant in human systems; and (e) flexibility and plasticity in development, body, and behavior are common (Fuentes 2009; Anton and Snodgrass 2012; Wells and Stock 2007).

Given this perspective, when we are talking about humans we need to situate our discussions in contexts that reflect a contemporary understanding of evolutionary processes integrated with diverse disciplinary perspectives on behavior, perceptions, institutions, and histories. Many of the contributions in this book are ripe for such a contextualization.

Putting Evolution to Work in Investigating Human Nature(s)

In the first chapter of this book, "Off Human Nature," we are challenged to move beyond seeing nature and culture and distinct entities and asked to envision a processes wherein culture and biology are simultaneously constructing and being constructed by the process of becoming human. In the first few chapters a set of twin narratives emerge: humans as biocultural creatures and humans as beings becoming themselves via entanglements with the social, perceptual, and material worlds around them. These concepts are rooted in an anthropological approach that envisions humans as enmeshed in webs of significance that are simultaneously biological, experiential, historical, and social (e.g., Geertz 1973). In this approach it makes no sense to talk about a fixed human nature or a "best way" to be human: the human experience is one of relational processes and becomings in a diversity of developmental and experiential contexts (Ingold 2004). And yet when connecting this perspective to a contemporary evolutionary context one is empowered to make some claims about human nature(s), and one can do so without having to retreat into a wholly reductionist or ultimately functional adaptationist explanatory framework.

The EES approach enables a broader template and playing field for involving a range of elements (perception, embodiment, history, institutions, etc.) in the conceptualization of how evolutionary processes are at play in the human experience. The possibilities that epigenetic, behavioral, and symbolic inheritances can share primacy with genetic inheritance in evolutionary processes provide a role for experience and perception as meaningful agents in the human niche. Therefore, the concepts of human becoming and the key dynamics of what it means to have culture become central aspects of the niche construction dynamic.

But does this move us any further in thinking about human nature(s)? Yes, it does. It provides us with a not necessarily reductive evolutionary context in which to place the moving target of the human. It enables us to construct meaningful scenarios about human evolutionary histories that include more than genetic adaptation as explanatory underpinnings and enables us to identify patterns in the dynamic evolutionary processes of human lives that are shared by members of our species. The EES approach enables us to see how particular evolutionary processes are components of human becoming, of human cultural experience, and can help us understand how, and why, we are biocultural ex-apes (Marks this volume). I suggest that it is in these synergies offered by melding the anthropological approaches outlined in chapters 1 and 2 with the affordances of the extended evolutionary synthesis that we can find central patterns and processes that offer fruitful locales for inquiry into what I would term "humans nature(s)."

The EES perspective I have outlined here also enables nonanthropologists and scholars outside the field of biology or the mechanics of evolutionary processes to participate in a discourse that involves evolutionary theory. If talking about evolution does not automatically limit one to a material reductive explanation and enables a space where human action, histories, institutions, and perceptions can have agency in evolutionary processes, then we can envision that beneficial engagements between different humanistic disciplines, such as theology and philosophy, and anthropologists and even biologists can occur. For example, one can see in the contributions in this volume by the theologians J. Wentzel van Huyssteen, Markus Mühling, Lluis Oviedo, and Neil Arner that while there are intellectual commitments where full overlap with anthropological and biological approaches cannot occur, the EES context offers a much broader landscape for possible interface in our communal attempts to develop a meaningful conversation about the human that is richly, and honestly, informed by multiple intellectual perspectives.

The EES also helps us situate Brown and Strawn's and Ramsey's contributions, which seek to provide content for a specific approach to identifying a, or patterns of a, human nature. Whether rooted in nonreductive physicalism or in biologically centered morphology and behavior, seeking to understand the neurobiology and social richness of the human niche and undertaking the quest for collections of traits that are central in the human experience can resonate with contemporary evolutionary theory.

However, such interfaces are not easily achieved. One cannot overlook that there are differences in vocabularies, philosophical commitments, and perceptual worldviews that permeate the essays in this volume. Even with the EES acting to provide some common ground for discussion, there remain the problems of mutual intelligibility and divergent basal beliefs and the near-impossibility of full concordance on certain ultimate accounts of the world. However, these discordances are not fatal blows to the endeavor. There is an emerging methodological approach that tackles such problematized discourse head-on: transdisciplinarity. In the final section of this essay I argue that taking a transdisciplinary approach might provide an effective way to navigate the complexities and divergences inherent in a discussion of human nature(s) that involves multiple disciplinary perspectives.

A Note on Transdisciplinarity and Moving Forward

In the introduction to this volume Aku Visala and I suggested that "the goal of transdisciplinarity is to change the disciplines involved by influencing the methods, worldviews, and languages used in each. This change goes deeper than just collaboration on similar subject matter as in interdisciplinarity and multidisciplinarity." This ideal is epitomized by the historian A. J. McMichael (2002, 220) when he states, "Transdisciplinarity is more than the mixing and interbreeding of disciplines. Transdisciplinarity transports us: we then ask different questions, we see further, and we perceive the complex world and its problems with new insights."

Interdisciplinary approaches focus on the spaces between disciplines in an effort to create particular connections that incorporate aspects of the assumptions, worldviews, and vocabularies of different disciplines (Kessel and Rosenfeld 2008). Multidisciplinary approaches, wherein individuals from different disciplines unite for an investigation, can produce novel or more comprehensive data but often lack the integrative framework, or shared lexicon, for analyses of those data. In both cases (inter- and multidisciplinary approaches) the possibilities of actual transformation of disciplinary boundaries or intellectual approaches is unlikely. With transdisciplinarity there is the goal of developing a relationship that facilitates the possibility of an intellectual transformation that is more thorough, intensive, and generative than in inter- or multidisciplinary approaches (Hadorn et al. 2008).

A transdisciplinary approach entails the utilization of a suite of factors as central parameters: rigor, openness, and tolerance (Kessel and Rosenfeld 2008). *Rigor* means taking into account as much existing information as is available in order to counteract the tendency for disciplinary biases and distortions to isolate potential explanatory frameworks from one another. This suggests that the disciplinary boundaries should not be the deciding factors in which "data" can be used in the analyses or discussion. *Openness* means a process of disciplinary generosity and an acceptance of potential unknowns as a positive context, not necessarily an obstacle. If participants from different disciplines and perspectives relax some of the strict assumptions about what constitutes "valid" approaches it can create a space for emergent and unexpected (even unforeseeable) outcomes between team members. This is not an argument that individuals should reject their own basal philosophical and methodological commitments; rather it is simply the possibility that they are able to entertain, for the purposes of the project, a (slightly) wider range of data and analytical lenses than is common for their specific disciplinary approach. *Tolerance*, also referred to as intellectual generosity, provides the space for collaborators to hold ideas and truths that may appear opposed to one another and still engage in a mutual process of discovery.

The essays and commentaries in this volume reflect a range of attempts to sincerely engage in rigor, openness, and tolerance in considering what we can, and do, say about human nature(s). They are a set of initial steps toward a transdisciplinary approach. But where do we go next?

In this essay I have suggested two possible practices for moving the conversation on human nature(s) forward: the use of the extended evolutionary synthesis as the core evolutionary paradigm and the deployment of transdisciplinarity. It is my assertion that taken together these two minor steps provide a platform that might enable a more sincere and successful suite of interchanges and analyses about human nature(s). The EES provides a more inclusive, and better supported, theoretical landscape for diverse approaches to seriously engage with evolutionary processes and contexts, and the transdisciplinary approach provides a methodological context wherein differences in philosophical and methodological commitments is not a de facto lethal blow to collaborative endeavors. The next steps are then to implement these suggestions. This is best done both in research projects that intentionally bring together two or more potentially divergent disciplines and in the continued practice of holding multidisciplinary conferences and workshops that assume rigor, openness, and tolerance as their basal modes of exchange. That is what I hope the readers of this volume are inspired to do.

Note

1. I should also point out that the EES is not favored by all evolutionary biologists and theorists. There remains contention about how to best conceptualize and integrate the current range of knowledge in regard to genomic, developmental, and selection processes and whether or not evolutionary theory does indeed need a "rethink." See the debate between Laland et al. and Wray et al. in the October 8, 2014, issue of *Nature*.

References

Andersson, C., A. Törnberg, and P. Törnberg. 2014. "An Evolutionary Developmental Approach to Cultural Evolution." *Current Anthropology* 55 (2): 154–74.

Anton, S. C., and J. Snodgrass. 2012. "Origin and Evolution of Genus *Homo*: New Perspectives." *Current Anthropology* 53 (6): 479–96.

Campbell, C., A. Fuentes, K. C. MacKinnon, S. Bearder, and R. Stumpf. 2011. *Primates in Perspective*. 2nd ed. New York: Oxford University Press.

Flynn, E. G., K. N. Laland, R. L. Kendal, and J. R. Kendal. 2013. "Developmental Niche Construction." *Developmental Science* 16 (2): 296–313.

Fuentes, A. 2009. "Re-Situating Anthropological Approaches to the Evolution of Human Behavior." *Anthropology Today* 25 (3): 12–17.

———. 2015. "Integrative Anthropology and the Human Niche: Toward a Contemporary Approach to Human Evolution." *American Anthropologist* 117 (2): 302–15.

Geertz, C. 1973. *The Interpretation of Cultures: Selected Essays by Clifford Geertz*. New York: Basic Books.

Hadorn, G. H., S. Biber-Klemm, W. Grossenbacher-Mansuy, H. Hoffmann-Riem, D. Joye, C. Pohl, U. Wiesmann, and E. Zemp. 2008. "The Emergence of Transdisciplinarity as a Form of Research." In *Handbook of Transdisciplinary Research*, ed. G. H. Hadorn et al., 19–42. Zurich: Springer.

Henrich, J. 2011. "A Cultural Species: How Culture Drove Human Evolution." *Psychological Science Agenda* 25 (11). www.apa.org/science/about/psa/2011/11/human-evolution.aspx.

Ingold, T. 2004. "Beyond Biology and Culture." *Social Anthropology* 12 (2): 209–21.

Kendal, J. 2012. "Cultural Niche Construction and Human Learning Environments: Investigating Sociocultural Perspectives." *Biological Theory* 6 (3): 241–50. doi:10.1007/s13752-012-0038-2.

Kessel, F., and P. L. Rosenfeld. 2008. "Toward Transdisciplinary Research: Historical and Contemporary Perspectives." *American Journal of Preventive Medicine* 35 (2S): S225–S234.

Jablonka, E., and M. Lamb. 2005. *Evolution in Four Dimensions: Genetic, Epigenetic, Behavioral, and Symbolic Variation in the History of Life.* Cambridge, MA: MIT Press.

Jablonka, E., and G. Raz. 2009. "Transgenerational Epigenetic Inheritance: Prevalence, Mechanisms, and Implications for the Study of Heredity and Evolution." *Quarterly Review of Biology* 84 (2): 131–76.

Laland, K. N., and G. Brown. 2011. *Sense and Nonsense: Evolutionary Perspectives on Human Behaviour.* 2nd ed. Oxford: Oxford University Press.

Laland, K. N., J. Kendall, and G. Brown. 2007. "The Niche Construction Perspective: Implications for Evolution and Human Behavior." *Evolutionary Psychology* 5: 51–66.

Laland, K. N., T. Uller, M. Feldman, K. Sterelny, G. Müller, A. Moczek, E. Jablonka, and J. Odling-Smee. 2014. "Does Evolutionary Theory Need a Rethink? Yes, Urgently." *Nature* 514: 161–64.

———. 2015. "The Extended Evolutionary Synthesis: Its Structure, Assumptions and Predictions." *Proceedings of the Royal Society B* 282. http://dx.doi .org/10.1098/rspb.2015.1019.

McMichael, A. J. 2002. "Assessing the Success or Failure of Transdisciplinarity." In *Transdisciplinarity: Recreating Integrated Knowledge,* ed. M. A. Somerville and D. J. Rapport, 218–22. Toronto: McGill-Queen's University Press.

Nowak, M. A., and R. Highfield. 2011. *SuperCooperators: Altruism, Evolution, and Why We Need Each Other to Succeed.* New York: Free Press.

O'Brien, M., and K. N. Laland. 2012. "Genes, Culture, and Agriculture: An Example of Human Niche Construction." *Current Anthropology* 53 (4): 434–40.

Pigliucci, M., and G. B. Muller. 2010. *Evolution: The Extended Synthesis.* Cambridge, MA: MIT Press.

Read, D. 2012. *How Culture Makes Us Human: Primate Evolution and the Formation of Human Societies.* Walnut Creek, CA: Left Coast Press.

Richerson, P., and R. Boyd. 2005. *Not by Genes Alone: How Culture Transformed Human Evolution.* Chicago: University of Chicago Press.

Sober, E., and D. S. Wilson. 1998. *Do Unto Others: The Evolution and Psychology of Unselfish Behavior.* Cambridge, MA: Harvard University Press.

Sterelny, K. 2012. *The Evolved Apprentice: How Evolution Made Humans Unique.* Cambridge, MA: MIT Press.

Strum, S. C. 2012. "Darwin's Monkey: Why Baboons Can't Become Human." *Yearbook of Physical Anthropology* 149 (55): 3–23.

Tomasello, M. 2014. *The Natural History of Human Thinking.* Cambridge, MA: Harvard University Press.

Wells, J. C. K., and J. T. Stock. 2007. "The Biology of the Colonizing Ape." *Yearbook of Physical Anthropology* 50: 191–222.

West-Eberhard, M. J. 2003. *Developmental Plasticity and Evolution.* New York: Oxford University Press.

Wilson, D. S., and E. O. Wilson. 2007. "Rethinking the Theoretical Foundation of Sociobiology." *Quarterly Review of Biology* 82 (4): 327–48.

MOVING US FORWARD?

CELIA DEANE-DRUMMOND

Why should we consider a *forward* movement when considering human natures? The term implies that once we consider the variety of perspectives as outlined in this volume there may be some tentative conclusions that can be reached about where intellectual discourse needs to go next. Certainly, as the editors, Agustín Fuentes and Aku Visala, argue in the introduction, philosophical divergence and even opposing views seem to be the order of the day, ranging from the extremes of Steven Pinker's (2002) evolutionary psychology that resists the view of the mind as a blank slate,[1] presupposed in social constructivism, to Jesse Prinz's (2012) argument ten years later that less attention should be paid to evolution and that culture and experience shape who we are as nurtured human beings. The prospect of a *universal* human nature or even an *innate* human nature is still popular among evolutionary psychologists such as Pinker. The basis for such a claim is claimed to be strictly a *biological* one.

Theologians entering this fray are likely to resist any biological basis for universality, which can seem too totalizing and deterministic, but they might be tempted to press for a different route for universal claims about human nature on an ontological basis, that is, guaranteed by divine fiat. Human rights, for example, are premised on the notion of a special and, arguably, *universal* belief in human dignity, bolstered through the concept of natural/human right that has since become secularized (Deane-Drummond 2015a). The relationship between the intelligibility emerging in natural law and that emerging in laws of nature is a particularly complex one, especially when *biological* laws of nature are considered (Schloss 2013). Further, as Jonathan Marks suggests, the way that biological laws have been co-opted for political narratives gives us reason to be pretty suspicious of too strong a biological basis for human nature. Tim Ingold is therefore correct to resist what amounts to essentialist biological views of human nature.

An emerging consensus among the anthropologists who have contributed to this volume resists an overemphasis on a sharp divide between biological and cultural interpretations of human nature(s). Among the theologians represented there is, perhaps, rather more divergence in methodological preference, ranging from the interdisciplinarity that puts primacy on the insights emerging from anthropological sciences, characteristic of J. Wentzel van Huyssteen, through to a much more explicit theological view of human nature parsed according to normative principles, as in Neil Arner's position. This reflects the degree to which different theologians are prepared to incorporate and give different relative weight to insights from the natural sciences, so any anti-naturalism as far as belief in *God* is concerned can quite legitimately either remain below the surface or become much more explicit in discussions of *human* nature. And it is equally possible for a theological ethicist to either espouse a form of naturalistic ethic or use normative principles emerging from traditional sources. In addition, there are those who lean toward continental philosophy or postmodern approaches, all of which tend to weaken traditional frameworks for thinking about human nature in theological terms. Markus Müling is somewhere between van Huyssteen and Arner, pressing for an ontological basis for human nature, derived from trinitarian theological principles, but at the same time taking account of evolutionary and scientific insights. Hence, while all three theologians in this volume offer interdisciplinary and perhaps even transdisciplinary perspectives, finding genuine shifts in understanding and the way each might be considered to be "moving forward" will vary depending on their starting points and presuppositions.

It is also interesting that both Fuentes and van Huyssteen cite each other's work in order to support their own views; for Fuentes, van Huyssteen is read as being *transdisciplinary*, while for van Huyssteen, Fuentes is certainly *interdisciplinary*. Van Huyssteen does not use the term *transdisciplinarity* in this volume, but he still cites Fuentes as supporting his transversal approach.

These shades of difference are, it seems to me, important, since *interdisciplinarity* means something different for Fuentes and my own work, namely, a close conversation from two or more different perspectives, whereas transdisciplinarity means a shift in perspective as a result of the interaction. *Transversality*, a term that van Huyssteen prefers, edges very close to transdisciplinarity, but, it seems to me, they are not identical.[2] Transdisciplinarity is the *fruit of transversal reasoning* when each discipline is shifting in a way that becomes self-conscious, not just interacting and engaging through bridge concepts that transversal reasoning presupposes.

Small, subtle differences become a little more pronounced once methodological issues are expressed in more practical ways. Van Huyssteen in his chapter in this volume gives high priority to the anthropological insights that emerge in evolutionary research on early hominins such as those inspired by the Neolithic excavations at Çatalhöyük in Turkey. So, citing a range of anthropological sources, he concludes, "language, self-awareness, imagination, consciousness, moral awareness, symbolic behavior, and mythology are probably the defining elements that really make us human." Nonetheless, huge challenges still relate to how such capabilities can leave their mark in the material record. For him, a post-foundationalist philosophy colors the way that theological insights can be used, which, in practice, means that any claims that theology might make necessarily shifts in the light of evolutionary anthropology once viewed through a transversal lens. Van Huyssteen's attention to imagination, especially religious imagination, and a metaphysics of transcendence certainly resonates strongly with Fuentes's understanding of human becoming and more specific questions about those features of *Homo sapiens sapiens* that were (are?) distinctive and that allowed this particular subspecies to be successful when others were not. Indeed, van Huyssteen is happy to press for the concept of *human uniqueness*, which again has a slightly different flavor when compared to human distinctiveness, since the latter implies that other extant hominin groups that were around at the same time as *H. sapiens* could well have had at least *some* of the characteristics that we find in *Homo sapiens sapiens*.

Van Huyssteen is correct to give a generalist perspective on religion in his analysis of its evolution in very deep history. However, I suggest that it is also by paying close attention to *specific religious traditions* that further insights about human nature are likely to be revealed that will then, in turn, lead to new scientific questions. Hence, if transdisciplinarity is to be meaningful in a practical sense it has to work both ways, so it needs to be willing to tease out the details of how different practices and knowledges are understood from very different metaphysical starting points and then allow particular questions to come to the surface that would not have been explored otherwise. And sometimes this comes from paying attention to the very ways in which two disciplines are not just the same, but actually very different.

Looking Back, Looking Forward?

One of the characteristics of theology that is distinct from the social sciences is that it is ready to draw on ancient historical texts as a way of illumination, insight, and creativity. Retrieval is therefore unabashed rather than considered outdated. While anthropologists may have an interest in historical anthropology, inasmuch as the discoveries of human remains in the material record and their interpretation is always ongoing, and a process of constant debate, discussion, and change, there is necessarily a perception of *moving forward* as far as that anthropological knowledge is concerned. Sometimes a visionary may have understood aspects that do not get sufficient attention until much later, but most of the time the sciences, including the social sciences, thrive on new discoveries or new interpretations. Niche construction is one of the most exciting new perspectives in evolutionary theory, as is the concept of entanglement between species, both of which can prompt different theological perceptions of the human, or rather allow traditions to be interpreted in different ways (Deane-Drummond 2014a, 2014b, 2015b; Deane-Drummond and Fuentes 2014). Once we look into deep evolutionary history, even the perspectives found in ancient theological texts start to seem very recent, but these texts are, I suggest, still able to offer insights regarding human nature, and theology, to be true to itself *as theology*, is concerned not just with God but also with a recollection and reinterpretation of these texts in the light of current experience.

Therefore, for the purposes of this commentary, I am deliberately going to *look back* to the work of a scholar who was hugely influential on Thomas Aquinas, whose own work has influenced, and continues to influence, present generations of theologians. In the theology faculty at Notre Dame, and even in the philosophy faculty, the number of scholars taking his work seriously crosses the divides of subdisciplines, from philosophical theology to moral theology to systematic theology to historical Christianity. So, what did his teacher, Albertus Magnus, say about human nature(s)? For Albertus, "the most outstanding human characteristic" is that

> only the human is a point of union between God and the world. For the human has in himself the divine intellect and through this he is sometimes elevated above the world to the extent that even the material of the world follows upon his thoughts. We see this in the best-born men who use their souls to bring about a transmutation of worldly bodies and are thus said to perform miracles. Moreover, even if that very part by which a human is bound to the world, he is not subordinated to it, but rather is set over it like a helmsman.[3]

It is the *religious aspect*, then, that is the most distinctive mark of humanity, according to this ancient writer, and the reported human capacity to effect unusual changes in the world, named here as miracles, seems to be connected with a porousness or human liminality with the divine. But this divinity in humanity is not just for spectacular displays, but rather to allow proper responsibility for the world, the role of *helmsman*; so it amounts to a sensitive steering rather than dominating capacity. In Tim Ingold's terms, it is not enough for humans just to have a bare life; to be a human the life needs to be fully and consciously *lived out* in a certain way. Further, if a human being resists that responsibility by becoming subordinate to the world, then, for Albertus, he is "discarding the honor of his humanity," taking on a characteristic of "a beast." So, through con-cupiscence the human becomes like a pig; through anger, a dog; through plunder, a lion; and "likewise for the other vices" (vol. 2, bk. 22, ch. 5). The human is therefore in one sense divine but in another sense beastly, and it is quite possible for human beings to move in both directions. Liminality works both ways, and humans rest somewhere in the border-land between animal and God. Albertus Magnus refers to humans and *other animals*, reminding us that humans are animals too. Some will find his human-animal comparative project offensive in a postmodern con-text, since those animals are always going to be "lower" in any hierarchy of being (Deane-Drummond and Clough 2009). But Albertus had no trouble considering that the ontological basis for not just vices but also many virtues are to be found in other than human worlds, and that is close to an ethnographic discussion of beastly morality.

Consider, for example, Albertus's discussion of the soul. He had no trouble thinking of the plant world as having a soul directed toward life, while that in the sentient animal world is directed toward sensing and that in the rational world toward understanding (vol. 2, bk. 21, ch. 1).[4] He believes, then, that the plant soul displays characteristics of temperance and chastity, according to its particular functions, while a sensitive (i.e., sensate animal) soul expresses humility, gentleness, fortitude, and many other virtues according to its particular desires or angers (vol. 2, bk. 21, ch. 1). And for him these same powers exist in a rational soul that humans possess. The "foundationalist" Catholic teaching that God somehow in-tervenes in human becoming in order to introduce the soul at conception, or even with the arrival of the first Adam and Eve, seems to be a long way from Albertus's nested interpretation of soul making, with its buildup to-ward what is distinctively human.

And here it is possible to trace an ambivalent aspect of his interpre-tation of human nature that has plagued thinkers for centuries. On the one hand, continuities between human beings and other animals are ac-knowledged, so that those characteristics found in them are also in us. On the other hand, humans are viewed as the most perfect beings, even if specific characteristics like sight are sharper in other creatures like eagles. According to Albertus, "The human is the most perfect animal not only by virtue of the addition of reason, but also in terms of the powers and manner of carrying out these powers, both sensible and rational" (vol. 2, bk. 21, ch. 1). This attention to human reason, to the mind, or the brain, as Carl Gillett puts it in this volume, has a legacy reaching back to the Middle Ages. So in spite of Albertus's two large volumes dedicated to the study of animals as a whole, in which the amount of text on the human is relatively small, such close attention to these other creatures seems ultimately to be for the sake of deepening understanding of what it means to be human, and humans are always, unless forfeited by an act of will, considered superior by their particular reasoning skill compared to all other creatures. And it is this bilateral legacy that seeps into the current philosophical debates that is evident in Pinker and Prinz, with Pinker focusing attention on human evolution, animality, and Prinz the constructive, cultural aspects of who we are as human beings.

One of the difficulties for any theologian entering the fray of current evolutionary anthropology is to find a way to express theologically the distinctive marks of what it is to be human without falling into the trap of naming those humans as somehow superior on account of those particular characteristics. It is clear that Albertus used the best science of his time to reach his philosophical and theological conclusions, some of which are not so far off the mark. For example, he was prepared to suggest that only humans have intellectual abilities that can supervene on other matters, so that irascible and concupiscent desires are, unlike those in other animals, "susceptible to the persuasion of reason,"[5] and "only the human has the power of speculation on intellectual theorums and of being delighted by a pleasurable thing that has no contrary" (vol. 2, bk. 22, ch. 5, §12). In fact, the power of reason that is capable of seeking universal truths was such a high priority in Albertus's thinking about human beings that he considered pygmies, at the time thought to be another species, less than human.[6]

But before we dismiss such texts entirely as politically offensive to modern readers on account of its seeming warrant not just for human exceptionalism but also for an implicit political racism, often lurking in the background of essentialist accounts of human nature(s), it is important to note a number of other threads in his work that remain relevant in thinking through what human nature is about. In the first place, he took account of the bodily aspects of human beings, so it is incorrect to view his work as essentialist for attending just to the intellectual and religious powers of humans. For example, it is only in book 22, *On the Natures of Animals*, that he discusses humans in any detail. There five chapters are devoted to humans, and of these, two are on the processes of reproduction and two are on different aspects of sexual intercourse, with only one chapter devoted to natural and divine properties. It was, therefore, an account that pays attention to bodiliness or creaturehood that has a totally different flavor from philosophical rationalistic theories that became dominant in the post-Kantian era. There are also books devoted respectively to birds (bk. 23; see below), "aquatic animals" (bk. 24), "serpents" (bk. 25), and "vermin" (bk. 25).

Albertus viewed certain psychological characteristics as unique to humans; for example, "the power of discerning the difference between what is honorable and what is base belongs to the human alone" (vol. 2, bk. 22, ch. 5, §11). Many scholars would still agree with that today. There are still questions that plague evolutionary anthropologists about the origin of shame and how this might be linked to morality, even if the basis of that shame varies between cultures and societies. This is still an active area of research, especially that related to conscience and morality, and more work could be done in this field (Boehm 2012). Albertus was also fascinated by the power of the human hand, which was used for purposes *beyond immediate needs*: it is the "organ of organs and the operative intellect," to be used in art (vol. 2, bk. 21, ch. 1, §5). And the thought immediately springs to mind, in the ubiquitous cave paintings depicting the hand were our early ancestors naming something that they had noted self-reflexively as different about being human?[7] Albertus reminds us not to forget about the hand in thinking about what it means to be human; it is about *practices* arising from those cognitive powers, which gives some clues that perhaps there will, after all, be proxies left behind in the archaeological record that can be tracked down and recorded.

Finally, before leaving Albertus's work it is worth remembering also that he was quite prepared to acknowledge those prudential capabilities of other animals that his contemporaries could easily find objectionable as they had a much sharper sense of the difference between humans and other creatures. He was especially impressed by the intelligence of birds, their ability to build nests, and what looked like creativity and even artistry.[8] Recognition of the special intelligence of birds, as well as primates, has become much more prominent in literature relatively recently (see, e.g., Clayton and Emery 2008). Albertus also compared the relative lack of intelligence in domesticated animals, such as sheep and cattle, with deer, which he believed showed a greater degree of initiative and prudence in the way they behaved, especially with their young, and also in the remarkable way that some wild and domesticated animals sought out antidotes in the event of poisoning, a behavior he believed had to have been learned by experience.[9] He even commented on the fact that a rat had been trained to hold a candle at the table of its master (vol. 2, bk. 21, ch. 3). Close observation of these differences reminds us that it is not enough, philosophically, just to think of humans in relation to animals as if all other animals are the same; rather they are active agents in their own right, shaping and forming their worlds in close entanglement with our own (Haraway 2003, 2008). Albertus may not have come up with the term *naturecultures*, but he was aware of those facets of natural animal behavior that seemed to imitate something similar to that occurring in human societies. His discussion of "marriage" in pigeons for what looked like exclusive pair bonding might seem quaint,[10] but at least he was more prepared than most of his contemporaries to risk using anthropomorphic language to further his understanding of what was happening in the social lives of animals. And it is just this kind of risky anthropomorphizing that has become fruitful in thinking about animal ethology (Horowitz and Bekoff 2007). Fuentes and many others are also prepared to use anthropological tools to investigate not just human societies but also the social lives of other animals in entanglement with our own, interspecies relationships, so we arrive at ethnoprimatology (Fuentes 2012) or even ethnoelephantology (Locke 2013).

Albertus also noticed the difference between the kind of industry found in bees, for example, and the cleverness in response to human instruction, as in domesticated animals, but he also paid attention to monkeys.[11] Among monkeys he was ready to acknowledge a "capacity for imagination and memory," but these are used in a way to draw on what is in immediate experience rather than to draw out a "purified universal," that is, "according as it is the principle of science, or art, or prudence or some other intellectual virtue" (vol. 2, bk. 21, ch. 3 §15). And to those who might doubt that a kind of "prudence or wisdom from intellect" could ever be present in other animals because "they do not have intellect" (vol. 2, bk. 8, ch. 1),[12] he had this to say: sentient animals participate in what he terms the "middle powers of the soul." This means that there are four other aspects that perfect the imaginative powers: (1) having an organ that is capable of receiving those images; (2) a purity and cleanness of spirit, so that monstrous or other distortions are not imagined; (3) a clear pathway from sense and common sense; and (4) a good disposition of other organs, such as the heart or stomach. Though his language is of course archaic, his belief in the many and varied factors that influence the imagination in humans and in other animals is entirely appropriate. So it is not the imagination *as such* that is lacking in other creatures, or characteristic only of humans, but rather the particular way that imagination is used and for what purpose and in what context.

This brings us back to the issue of a sense of the transcendent, arguably only present in a self-conscious way in humans, even if some advanced social animals could, at a stretch, be thought to have a capacity for wonder. But how could a transcendent sense ever be mapped or measured in deep history? Human wisdom is associated with centuries of religious reflection, which is characteristic of virtually all religions, from Confucianism to Buddhism to monotheistic faiths. The literature on religious aspects of wisdom is vast, but all wisdom literature relates in an important way to the role of an individual in social and community functions. It is involved, therefore, in the particular creativity of sociality and relationships of mutual creation. Ingold terms this "the verb" of human being in the world, whereas theologically I prefer the language of performance (Deane-Drummond 2014a, 2014b), since it connotes a sense of direction in that sociality, though the two ideas are close. What Albertus recognized centuries ago is that the kind of sociality of wisdom found in human beings is distinctive; it reaches out to the transcendent and even, in experience, receives its power from the divine. We are liminal creatures, set between animality and divinity, yet also horizontally, as it were, stretched out in entangled communities, being shaped even in our moral worlds by interspecies interactions as well as shaping those other animal societies (Deane-Drummond 2014b; Deane-Drummond and Fuentes 2014). We have known for some time that such interactions are important in well-functioning ecological networks; but what has not been so clear is that they are also built into deep history. And that means that even wisdom and those specific characteristics that we think of as marking our human distinctiveness have taken shape from interaction between species. A project being undertaken at Notre Dame from 2015 to 2018 is the search for traces of that wisdom in the archaeological record.[13] How are we to find proxies for what is an internal state that expresses itself in community relations, yet reaches out to the transcendent? These are puzzles that spur us to think in new and different ways about what it means to be human, that creature doubly imbibed with wisdom (Ingold 2010).

Notes

1. Somewhat confusingly in my view, Steven Pinker's title *The Blank Slate* is the very view he seeks to oppose. His tendency to caricature his opponents has been heavily criticized.

2. In van Huyssteen's own words in this volume, "In the dialogue between theology and other disciplines, transversal reasoning promotes different, nonhierarchical but equally legitimate ways of viewing specific topics, problems, traditions, or disciplines, and creates the kind of space where different voices need not always be in contradiction, or in danger of assimilating one another, but are in fact dynamically interactive with one another."

3. Vol. 2, bk. 22, tr. 1, ch. 5, "On the Natural and Divine Properties of the Human," §9.

4. Bk. 21, *Which Is on Perfect and Imperfect Animals and the Reason for Their Perfection and Imperfection*, tr. 1, On the Degrees of Perfect and Imperfect, ch. 1, "On the Highest Perfection of Animal Which is the Human," §2.

5. Bk. 20, *On the Nature of Animal Bodies*, tr. 2, On the Formal Powers, ch. 6, "How Animals Differ One to the Other and How the Human Differs from Them All," §88.

6. Bk. 21, tr. 1, ch. 2, §11. He failed, like others at the time, to recognize the creativity in other cultures, and in this respect his view is limited by its cultural specificity.

7. This common image found in cave art is well represented in the Chevaux cave in France, dating from about 31,000 years ago.

8. He devotes more than one book to exploring different birds, dividing them according to size; his view is that the smallest birds are the most intelligent and that swallows are particularly clever. Vol 1, bk. 8, *On Animal Habits*, tr. 2, On Animal Prudence and Stupidity, ch. 3, §52.

9. For discussion of deer, see vol. 1, bk. 8, *On Animal Habits*, tr. 2, ch. 1; for animals seeking herbal remedies, see ch. 2.

10. Vol. 1, bk. 8, tr. 2, On Animal Prudence and Stupidity, ch. 3, §54.

11. He dedicates a whole chapter to monkeys: Vol. 2, bk. 21, tr. 1, ch. 3, "How Animals are Capable of Instruction by Some Participation in the Virtues of the Soul and Especially How This Occurs in the Genuses of Monkeys."

12. Vol. 1, bk. 8, tr. 6, ch. 1, "Animal Prudence and Stupidity."

13. Agustín Fuentes and I have received funding from the John Templeton Foundation for a project on human distinctiveness, which includes a core project on the evolution of wisdom. The definition of what that wisdom might be and how to search for appropriate proxies in the archaeological record is still being worked out as this book goes to press.

References

Albertus Magnus. 1999. *On Animals: A Medieval Summa Zoologica*. 2 vols. Trans. K. F. Kitchell and I. M. Resnick. Baltimore, MD: Johns Hopkins University Press.

Boehm, C. 2012. *Moral Origins: The Evolution of Virtue, Altruism, and Shame*. New York: Basic Books.

Clayton, N. S., and N. J. Emery. 2008. "Canny Corvids and Political Primates: A Case for Convergent Evolution in Intelligence." In *The Deep Structure of Biology: Is Convergence Sufficiently Ubiquitous to Give a Directional Signal?*, ed. S. C. Morris, 128–42. Conshohocken, PA: Templeton Foundation Press.

Deane-Drummond, C. 2014a. "Evolutionary Perspectives on Inter-Morality and Inter-Species Relationships Interrogated in the Light of the Rise and Fall of *Homo sapiens sapiens*." *Journal of Moral Theology* 3 (2): 72–92.

———. 2014b. *The Wisdom of the Liminal: Human Nature Evolution and Other Animals*. Grand Rapids, MI: Eerdmans.

———. 2015a. "Natural Law Revisited: Wild Justice and Human Obligations to Other Animals." Paper presented at the Society for Christian Ethics, Chicago, January 11.

———. 2015b. *Re-Imaging the Divine Image: Humans and Other Animals*. Goshen Lectures. Kitchener, ON: Pandora Press.

Deane-Drummond, C., and D. Clough, eds. 2009. *Creaturely Theology: On God, Humans and Other Animals*. London: SCM Press.

Deane-Drummond, C., and A. Fuentes. Forthcoming. "Human Being and Becoming: Situating Theological Anthropology in Interspecies Relationships in an Evolutionary Context." *Philosophy, Theology and the Sciences* 1 (2).

Fuentes, A. 2012. "Ethnoprimatology and the Anthropology of the Human-Primate Interface." *Annual Review of Anthropology* 41: 101–17.

Haraway, D. 2003. *The Companion Species Manifesto: Dogs, People, and Significant Others*. Chicago: Prickly Paradigm.

———. 2008. *When Species Meet.* Minneapolis: University of Minnesota Press.

Horowitz, A., and M. Bekoff. 2007. "Naturalizing Anthropomorphism: Behavioral Prompts to Our Humanising of Animals." *Anthrozoos: A Multidisciplinary Journal of the Interactions of People and Animals* 20: 23–35.

Locke, P. 2013. "Explorations in Ethnoelephantology: Social, Historical, and Ecological Intersections between Asian Elephants and Humans." *Environment and Society: Advances in Research* 4: 79–97.

Pinker, S. 2002. *The Blank Slate: The Modern Denial of Human Nature.* New York: Viking.

Prinz, J. 2012. *Beyond Human Nature: How Culture and Experience Shape the Human Mind.* New York: W. W. Norton.

Schloss, J. 2013. "Laws of Life." In *Concepts of Law in the Sciences*, ed. M. Welker and G. Etzelmüller, 61–82. Legal Studies and Theology, Religion in Philosophy and Theology, 72. Tübingen: Mohr Siebeck.

CONTRIBUTORS

NEIL ARNER is Assistant Professor at the University of Notre Dame. His research focuses on the integration of natural law and divine command forms of Christian ethics, the contemporary theological relevance of early modern thought, the prospects for a Protestant recovery of natural law theory, the potential for ecumenical and interreligious collaboration in addressing common moral concerns, and theological responses to scientific studies of the "origins" of morality. Arner teaches an undergraduate University Seminar on God's reign in the biblical narrative as well as graduate-level courses on evolutionary ethics, ecumenical ethics, the history of Christian ethics, and methods in fundamental moral theology.

SUSAN D. BLUM is Professor of Anthropology at the University of Notre Dame. She is interested in questions of human experience through cultural formations such as nationalism, ethnicity, linguistic interaction, and education. Her most recent publication is *"I Love Learning; I Hate School": An Anthropology of College* (Cornell 2016).

WARREN BROWN is Director of the Travis Research Institute and Professor of Psychology at Fuller Theological Seminary. Currently, he is involved in neuroscience research related to the cognitive and psychosocial disabilities in a congenital brain malformation called agenesis of the corpus callosum. Brown has also studied callosal function in dyslexia, attention deficit hyperactivity disorder, multiple sclerosis, and Alzheimer's disease, and has conducted research on brain wave changes associated with aging and dementia, language comprehension, dialysis treatment for kidney disease, and attention deficits in schizophrenia. Most recently, Brown and colleagues from other institutions have been involved in research on the psychology and neuroscience of exemplars of the virtues of compassion and generosity.

JAMES M. CALCAGNO is Professor of Anthropology and Director of the Fellowship Office at Loyola University Chicago. He is a biological anthropologist whose research interests have ranged from the mechanisms of human dental evolution to primate behavior to the most basic question of anthropology, What makes us human? He has conducted research in Europe, Africa, and the United States and has published in the *American Journal of Physical Anthropology, American Anthropologist, Journal of Human Evolution, Evolutionary Anthropology*, and *Zoo Biology*. Most recently, he has been involved in an online course on what makes us human from an evolutionary perspective through TEDTalks.

KELLY JAMES CLARK is Professor in the Honors Program at Brooks College and Senior Research Fellow at the Kaufman Interfaith Institute at Grand Valley State University in Grand Rapids, Michigan. His interests are religious liberty and tolerance, philosophy of religion, ethics, science and religion, and Chinese thought and culture. His books include *Religion and the Sciences of Origins, Return to Reason, The Story of Ethics, When Faith Is Not Enough*, and *101 Key Philosophical Terms and Their Importance for Theology*. Clark's *Philosophers Who Believe* was voted one of *Christianity Today's* 1995 Books of the Year.

CELIA DEANE-DRUMMOND is Professor of Theology and Director of the Center for Theology, Science and Human Flourishing at the University of Notre Dame. Her work focuses on the intersection of theology and ethics with the biological sciences, in particular genetics, ecology, anthropology, and ethology. Deane-Drummond has been chair of the European Forum for the Study of Religion and Environment (EFSRE) since 2011 and has collaborated extensively with the Catholic Fund for Overseas Development (CAFOD) focusing on climate change and environmental justice. She is currently joint editor of the new journal *Philosophy, Theology and the Sciences*.

AGUSTÍN FUENTES is Professor and Chair of Anthropology at the University of Notre Dame. He received a BA in zoology and anthropology and an MA and PhD in anthropology at the University of California, Berkeley. His current research focuses on cooperation, community, and semiosis in human evolution; ethnoprimatology and multispecies anthropology; evolutionary theory; and interdisciplinary approaches to human nature(s).

CARL GILLETT is Professor of Philosophy at Northern Illinois University. His research areas are the philosophy of mind/psychology, the philosophy of science, and metaphysics, and he is also interested in the philosophy of neuroscience and philosophy of religion. He has published in all of these areas in a variety of journals, including *Analysis*, the *Journal of Philosophy*, *Faith and Philosophy*, and *Nous*. In the spring semester of 2015, he was a fellow of the Durham Emergence Project, working on emergence in cognitive neuroscience, which brings together his earlier work on emergence/reduction in the sciences with his new project on the foundations of neuroscience and human nature.

DOUGLAS HEDLEY is Reader in Hermeneutics and Metaphysics at Cambridge and Co-Chair of the Platonism and Neoplatonism section of the American Academy of Religion. He is past secretary of the British Society for the Philosophy of Religion and past president of the European Society for the Philosophy of Religion. Hedley's research interests include contemporary philosophy of religion, the history of Platonism-Neoplatonism, early modern philosophy, and romanticism and idealism.

TIM INGOLD is Professor of Social Anthropology at the University of Aberdeen and Fellow of the British Academy and the Royal Society of Edinburgh. He has carried out fieldwork in Lapland and has written on the comparative anthropology of the circumpolar North, evolutionary theory, human-animal relations, language and tool use, environmental perception, and skilled practice. His current work explores the interface between anthropology, archaeology, art, and architecture. In 2014 Ingold was honored with the Royal Anthropological Institute's Huxley Memorial Medal.

IAN KUIJT is Professor of Anthropology at the University of Notre Dame and an anthropological archaeologist interested in the materiality of social differentiation, human agency, and evolution of social inequality. His research focuses on archaeology and the history of village life and the household, the origins of Neolithic agriculture and inequality, the archaeology and history of post-seventeenth-century Ireland, the emergence of social differentiation and the materialization of identity, and the prehistory of western North America.

JONATHAN MARKS is Professor of Anthropology at the University of North Carolina at Charlotte. His primary training is in biological anthropology and genetics, but his interests are broad. His work has received the W. W. Howells Book Prize and the General Anthropology Division Prize for Exemplary Cross-Field Scholarship from the American Anthropological Association and the J. I. Staley Prize from the School for Advanced Research.

MARKUS MÜHLING is Professor of Systematic Theology and International-Interdisciplinary Dialogue at the Leuphana University in Lüneburg, Germany. He previously taught at the University of Heidelberg, Germany, and King's College, University of Aberdeen, Scotland. His research interests are questions of systematic theology, such as trinitarian theology, atonement, eschatology, and ethics; and the dialogue between theology and science.

DARCIA NARVAEZ is Professor of Psychology at the University of Notre Dame. Her research explores questions of moral cognition and moral development over the life span in multiple contexts, for example, family and school. She examines how early life experience influences moral functioning and moral character in children and adults, integrating neurobiological, clinical, developmental, and education sciences in her theories and research about moral development. Questions that interest her include how early experience shapes moral capacities; what types of moral orientations people can have; how moral dispositions develop; what is moral wisdom and how we cultivate it; and how educators and parents can foster optimal development, well-being, and communal imagination.

LLUIS OVIEDO is Professor of Theological Anthropology at the Pontifical University Antonianum of Rome and Professor of Fundamental Theology at the Theological Institute of Murcia (Spain). Oviedo currently leads a project titled "Interdisciplinary and Empirical Theology" at the Antonianum University, which focuses on dialogue between theology and the human and social sciences, and is a team member of the Human Specificity project, funded by the Templeton Foundation. He previously served as editor of the periodical *Antonianum* (1997–2004) and is currently editor of the bulletin of the European Society for the Study of Science and Theology, *ESSSAT News & Reviews*.

GRANT RAMSEY is BOFZAP Research Professor in the Institute of Philosophy at KU Leuven, Belgium. His work centers on the philosophical problems at the foundation of evolutionary biology. He has published widely in this area, as well as in the philosophy of animal behavior, human nature, and the moral emotions. He runs the Ramsey Lab, a highly collaborative research group focused on issues in the philosophy of the life sciences.

PHILLIP R. SLOAN is Professor Emeritus of the Program of Liberal Studies and the Program of History and Philosophy of Science at the University of Notre Dame. His scholarly work has focused on the history and philosophy of life science from the Enlightenment period to the present. He is an editor of *Darwin in the Twenty-First Century: Nature, Humanity, and God* (2015). His current project, *Mastering Life,* is an analysis of the concept of life in modern biology and its ethical implications.

RICHARD SOSIS is James Barnett Professor of Humanistic Anthropology and Director of the Evolution, Cognition, and Culture Program at the University of Connecticut. His work has focused on the evolution of religion and cooperation, with particular interests in ritual, magic, religious cognition, and the dynamics of religious systems. To explore these issues, he has conducted fieldwork with remote cooperative fishers in the Federated States of Micronesia and with various communities throughout Israel. He is cofounder and coeditor of the journal *Religion, Brain & Behavior,* which publishes research on the biological study of religion.

BRAD D. STRAWN is Evelyn and Frank Freed Professor of the Integration of Psychology and Theology and Chair of Integration, Department of Clinical Psychology, School of Psychology, at the Fuller Theological Seminary. His areas of expertise are psychology and Christian theology, psychoanalytic psychotherapy, clinical psychology, and body, brain, and theology. Strawn is coauthor, with E. Bland, of *Christianity and Psychoanalysis: A New Conversation* (2014); he has also published numerous articles in journals, collections, and dictionaries.

LINDA SUSSMAN is Research Associate in the Department of Anthropology at Washington University, St. Louis. She is a medical anthropologist and epidemiologist whose research has focused on traditional medical beliefs and practices, ethnobotany, interpretations of illness, treatment-seeking and health care decisions, and how they relate to cultural factors such as ethnic identity, religion/worldview, social structure and organization, and the economic system. She has conducted fieldwork in southwestern Madagascar, on the polyethnic island of Mauritius, and in low-income urban areas of the United States. Sussman has published in a variety of journals including *Social Science and Medicine, Journal of Ethnopharmacology,* and *Pediatrics* and has contributed to textbooks on understanding the importance of culture in psychological counseling and the treatment of chronic conditions in multiethnic societies.

ROBERT SUSSMAN was Professor in the Department of Anthropology, Washington University, St. Louis, Fellow of the American Association for the Advancement of Science, and Chair of Section H (Anthropology) of the AAAS. He was a cofounder and the first president of the Midwest Primate Interest Group and past editor of the *Yearbook* of the *American*

Journal of Physical Anthropology. Along with his coauthor, Donna Hart, Sussman was the recipient of the W. W. Howells Book Prize in 2006 for the volume *Man the Hunted: Primates, Predators, and Human Evolution*. He was past editor in chief of *American Anthropologist* and past secretary, Section H (Anthropology), AAAS.

J. WENTZEL VAN HUYSSTEEN is James I. McCord Professor of Theology and Science Emeritus at the Princeton Theological Seminary. His primary research focus has been interdisciplinary theology, specifically theology and science/philosophy of science. Van Huyssteen has received a number of awards over his career, including the Andrew Murray Prize for Theological Literature, the Bill Venter Award for Academic Excellence, and the American Academy of Religion Senior Research Award. He was the first South African and the first Princeton Theological Seminary professor to be invited to deliver the acclaimed Gifford Lectures at the University of Edinburgh in 2004. He was also the first recipient of the Andrew Murray–Desmond Tutu Prize in South Africa, awarded in 2007, and was granted an Honorary Doctorate of Theology in Systematic Theology from Stellenbosch University in 2009. His work has included research in theological methodology, anthropology, philosophical theology, exploring such themes as how religious faith relates to culture and science, and paleoanthropology.

AKU VISALA is University Researcher at the Centre of Excellence in Reason and Religious Recognition Research, Faculty of Theology, at the University of Helsinki. He holds a PhD in the philosophy of religion from the University of Helsinki, Finland. He has previously held research positions in the Department of Anthropology at the University of Notre Dame, at the Centre of Anthropology and Mind at the University of Oxford, and the Center for Theological Inquiry in Princeton. His current research interests include theological and philosophical anthropologies as they relate to the cognitive sciences and evolutionary psychology.

INDEX

abstract thought, 63
acquisition, 250
adaptation
  cultural, 28–29, 52
  genetic, 253
  human, 115
aesthetic preferences, 222
affective neuroscience, 156
'Ain Ghazal, 193
Albertus Magnus, 264–67
Alone in the World? (van Huyssteen),
  170, 212
altruism, 6, 217
Ambady, Nalini, 45
American Anthropological Association, 27
ancestry vs. identity, 51–52
Andersson, C., 251
Angell, T., 60
animal culture, 60–62, 64–65, 111n3,
  268
animalism, 18–19, 73, 155, 160–62
anti-essentialism, 14–15, 240–43. *See
  also* essentialism
anti-naturalism, 261
apes, 34
  ancestry, 28
  culture, 45
Aquinas, Thomas, 205

Aristotelian human nature, 52–53
Aristotle, 208
attachment style, 134–35
attentionality, 81–82
Aureli, F., 61
auto-fabricators, humans as, 75

becoming human, 170, 171, 180,
  185n7. *See also* human becoming;
  human beings
behavior, human
  evolution in, 169–70
  looking for explanations of, 8
  pattern variations, 29–30
behavior patterns, 61–62
being human
  becoming human vs., 170
  defined, 63, 185n7
  transition between becoming
    human and, 171, 180
  See also human becoming; human
    beings
Bergson, Henri, 77–78
Beyond Human Nature (Prinz), 242
bias, 143
binary oppositions, 30
biocultural nature, 27–28, 42–43,
  202–3
biological evolution. *See* evolution

CEDLRN (*cont.*)
  emergence of, 129–30
  neural openness, 130–31
  neurophysiologicalism, 126–27
concrescence, 77
concupiscence, 219
condilectus, 109
Confucians, 90
connectome, 156
consciousness, 141, 142–43, 262
constitutionalism, 19
constructivism, 11–12, 53–54, 109,
    168–69, 184n2, 201–2
contradiction, living with, 99
corvids, 60
Cosmides, Leda, 7–8
Creative Evolution (Bergson), 77
creative good, 76–77
creativity, 76–77, 100
critical scientific realism, 11
cultural neuroscience, 45–46
cultural patterns, 252
culture
  anthropological concept of, 59
  constructing nature, 30–32
  defined, 65–66
  differences in, 59
  human, 207–8
  human nature and, 62–64, 65–67
  as intrinsic part of human condi-
    tion, 30–32
  multigenerational benign, 61
  nature vs., 227
  nature without, 32–35, 46–47
  in nonhuman primates, 60–61
  pre-Aristotelian concept of, 33
Curtis, Susan, 132

Daoism, 89–90
Darwin, Charles, 5–6, 50–51, 223

Darwin Deleted (Bowler), 50
Davenport, Charles B., 58
Deacon, Terrence, 132, 176
Dennett, Daniel, 208
deontological ethics, 18
deprivation, 135–36
descent with modification, 27
desideratum, 243
determinism, 151
de Waal, F. B. M., 62
Diamond, Jared, 35
direct perception theory, 117
distinctiveness, human, 167, 174–76,
    262–63
divinity in humanity, 264–65
dolus eventualis, 200
dominant symbolism, 196
Douglas, John H., 71
dualism
  argument against, 150
  consequences of, 152–53
  creation in God's image, 150
  defined, 19, 124–25
  evolution and, 153
  folk, 150
  free will and, 151
  geocentrism and, 150
  self-organizing personhood and,
    145
  substance, 150–51
  support for, 150–52
  in Western Christian thought,
    148–49
dual process theories, 142–43
Duple (word game), 96
Dupré, John, 241
duration, time as, 78
dynamic systems theory, 128–30, 250
dysnarrativia, 133

as evolutionary force, 169, 251–52
human nature as, 64–65
Marx and, 42
niche reception, 109
nobility of sight, 209
nonessentialists, 224–25
nonhuman animals, 60–62
nonhuman species, 244
nonreductive physicalism, 124
nonverbal communication, 105
normative concept of human nature
aesthetic preferences, 222
denial of, 53
descriptive accounts vs., 222
as guide to action, 217–18
metaphysical approach to, 234–37
natural selection and, 221–22
religious in origin, 218–20
normativity, 234–37
notion of transversality, 185n9
nouns, use of, as dispositional terms,
97
nurture, 45–46, 241–42

O'Brien, M., 251
observation, 80–81
ontogenesis, 77–78
openness, 256
opportunity teaching, 60
organism-environment interaction,
12, 250
original sin, 219
Origin of Species (Darwin), 223
Ortega y Gasset, José, 74–75, 78

Paloutzian, R. F., 144
Panksepp, Jaak, 156, 159
Park, C. L., 144
path of observation, 80–81

pathos, 108–9
patterns
behavioral, 29–30, 61–62
of constraint, 129
cultural, 252
extended evolutionary synthesis
(EES) and, 254
of interactivity, 128
self-organizing dynamic, 130
variations, 29–30
perception, 118–19
perceptual attunement theory, 81
permaculture, 91
person, defined, 105
personal identity, 142
personality, human, 104–5
personhood
alternative forms of, 125
complexity, 127–29
defined, 105–6
dualism and, 145
embodiment of, 125
good and evil experience, 178–79
identity and, 190
imago Dei and, 178
language development, 131–33
Levantine Neolithic, 191
material identity and, 196–97
Neolithic face and, 192–94
neurophysiologicalism, 126–27
production of, 166
relationality of, 133–36
relationships and, 100–101
self-organizing nature of, 131,
140–46
social construction of, 191–92
storytelling, 133
persons
as becomings, 104–5